RealTime Physics

Active Learning Laboratories

Module 1

Mechanics

A cheetah can accelerate from 0 to 50 miles per hour in 6.4 seconds.
—Encyclopedia of the Animal World

A Jaguar can accelerate from 0 to 50 miles per hour in 6.1 seconds.
—World Cars

David R. Sokoloff
Department of Physics
University of Oregon

Ronald K. Thornton
Center for Science and Math Teaching
Departments of Physics and Education
Tufts University

Priscilla W. Laws
Department of Physics
Dickinson College

John Wiley & Sons, Inc.

ACQUISITIONS EDITOR	Stuart Johnson
PUBLISHER	Kaye Pace
PROJECT EDITOR	Geraldine Osnato
MARKETING MANAGER	Robert Smith
PRODUCTION DIRECTOR	Pamela Kennedy
PRODUCTION EDITOR	Sarah Wolfman-Robichaud
SENIOR DESIGNER	Kevin Murphy
ILLUSTRATION EDITOR	David Renwanz

This book was set in Palatino by The GTS Companies/York, PA Campus and printed and bound by Courier/Westford. The cover was printed by Phoenix Color.

This book is printed on acid-free paper.

Printed in the United States of America

10 9 8 7 6 5

Preface

Development of the series of *RealTime Physics (RTP)* laboratory guides began in 1992 as part of an ongoing effort to create high-quality curricular materials, computer tools, and apparatus for introductory physics teaching.[1] The *RTP* series is part of a suite of *Activity-Based Physics* curricular materials that include the *Tools for Scientific Thinking* laboratory modules,[2] the *Workshop Physics* Activity Guide,[3,4] and the *Interactive Lecture Demonstration* series.[5] The development of all of these curricular materials has been guided by the outcomes of physics education research. This research has led us to believe that students can learn vital physics concepts and investigative skills more effectively through guided activities that are enhanced by the use of powerful microcomputer-based laboratory (MBL) tools.

In the past twelve years new MBL tools—originally developed at Technical Education Research Centers (TERC) and at the Center for Science and Mathematics Teaching, Tufts University—have become increasingly popular for the real-time collection, display, and analysis of data in the introductory laboratory. MBL tools consist of electronic sensors, a microcomputer interface, and software for data collection and analysis. Sensors are now available for motion, force, sound, magnetic field, current, voltage, temperature, pressure, rotary motion, acceleration, humidity, light intensity, pH, and dissolved oxygen.

MBL tools provide a powerful way for students to learn physics concepts. For example, students who walk in front of an ultrasonic motion sensor while the software displays position, velocity, or acceleration in real time more easily discover and understand motion concepts. They can see a cooling curve displayed instantly when a temperature sensor is plunged into ice water, or they can sing into a microphone and see a pressure vs. time plot of sound intensity.

MBL data can also be analyzed quantitatively. Students can obtain basic statistics for all or a selected subset of the collected data and then either fit or model the data with an analytic function. They can also integrate, differentiate, or display a fast Fourier transform of data. Software features enable students to generate and display *calculated quantities* from collected data in real time. For example, since mechanical energy depends on mass, position, and velocity, the time variation of potential and kinetic energy of an object can be displayed graphically in real time. The user just needs to enter the mass of the object and the appropriate energy equations ahead of time.

The use of MBL tools for both conceptual and quantitative activities, when coupled with recent developments in physics education research, has led us to expand our view of how the introductory physics laboratory can be redesigned to help students learn physics more effectively.

COMMON ELEMENTS IN THE *REALTIME PHYSICS* SERIES

Each laboratory guide includes activities for use in a series of related laboratory sessions that span an entire quarter or semester. Lab activities and homework assignments are integrated so that they depend on learning that has occurred during the previous lab session and also prepare students for activities in the next session. The major goals of the *RealTime Physics* project are: (1) to help students acquire an understanding of a set of related physics concepts; (2) to provide students with direct experience of the physical world by using MBL tools for real-time data collection, display and analysis; (3) to enhance traditional laboratory skills; and (4) to reinforce topics covered in lectures and readings using a combination of conceptual activities and quantitative experiments.

To achieve these goals we have used the following design principles for each module based on educational research.

- The materials for the weekly laboratory sessions are sequenced to provide students with a coherent observational basis for understanding a single topic area in one semester or quarter of laboratory sessions;
- The laboratory activities invite students to construct their own models of physical phenomena based on observations and experiments;
- The activities are designed to help students modify common preconceptions about physical phenomena that make it difficult for them to understand essential physics principles;
- The activities are designed to work best when performed in collaborative groups of 2 to 4 students;
- MBL tools are used by students to collect and graph data in real time so they can test their predictions immediately;
- A learning cycle is incorporated into each set of related activities that consists of prediction, observation, comparison, analysis, and quantitative experimentation;
- Opportunities are provided for class discussion of student ideas and findings;
- Each laboratory includes a pre-lab warm-up assignment, and a post-lab homework assignment that reinforces critical physics concepts and investigative skills.

The core activities for each laboratory session are designed to be completed in two hours. Extensions have been developed to provide more in-depth coverage when longer lab periods are available. The materials in each laboratory guide are comprehensive enough that students can use them effectively even in settings where instructors and teaching assistants have minimal experience with the curricular materials.

The curriculum has been designed for distribution in electronic format. This allows instructors to make local modifications and reprint those portions of the materials that are suitable for their programs. The *Activity-Based Physics* curricular materials can be combined in various ways to meet the needs of students and instructors in different learning environments. The *RealTime Physics* laboratory guides are designed as the basis for a complete introductory physics laboratory program at colleges and universities. But they can also be used as the central component of a high school physics course. In a setting where formal lectures are given, we recommend that the *RTP* laboratories be used in conjunction with *Interactive Lecture Demonstrations.*

THE MECHANICS LABORATORY GUIDE

The primary goal of this *RealTime Physics Mechanics* guide is to help students achieve a solid understanding of classical mechanics including Newton's three laws of motion. A number of physics education researchers have documented that most students begin their study of mechanics with conceptions about the nature of motion that can inhibit their learning of Newton's Laws.[6]

Newtonian dynamics is basically a study of the relationship between force and motion. The simultaneous use of an MBL force probe and motion sensor is powerful because students can display force-time graphs in real time along with any combination of graphs of position, velocity and/or acceleration versus time. The availability of a low-friction dynamics cart and track system makes possible observations of forces and the resulting motions initially in the simplest possible cases.

A critical aspect of *RealTime Physics Mechanics* is the use of a different order of presentation of topics than that used in popular textbooks. This order of topics, which we have dubbed the *New Mechanics* sequence, is based on our own experience with class testing and evaluation of the *Tools for Scientific Thinking* and *Workshop Physics* curricula. Based on earlier suggestions by Rothman[7] and Arons,[8] we have re-sequenced activites in both *RealTime Physics* and *Workshop Physics* to treat one-dimensional kinematics and dynamics before any two-dimensional motions are considered, and to introduce linear momentum before energy. The *New Mechanics Sequence* is described in more detail in a recent article.[9] The key elements are:

- One-dimensional kinematics and dynamics are treated before two-dimensional situations.
- Newton's Second Law is developed before Newton's First Law.
- The relationship between visible applied forces and motions in situations where friction is negligible is treated before the consideration of other types of forces.
- The "invention" of invisible forces such as sliding friction and gravity is encouraged as a way to preserve the viability of the Newtonian Schema for predicting motions.
- Newton's Third Law is developed through direct observations of the interaction forces between objects both during short-duration collisions and when objects are in contact for longer times.
- Momentum concepts are treated before the introduction of mechanical energy concepts.

RealTime Physics Mechanics includes 12 labs covering kinematics, dynamics, gravitational forces, passive forces, projectile motion, impulse and momentum, and work and energy.

Labs 1 and 2 (Introduction to Motion and **Changing Motion):** These labs involve the study of one-dimensional kinematics using an MBL system with a motion sensor. After recording their own body motions, students use a battery-operated fan unit mounted on a low-friction dynamics cart to produce constant accelerations.

Labs 3 through 7 (Force and Motion, Combining Forces, Force, Mass and Acceleration, Gravitational Forces, and **Passive Forces and Newton's Laws):** These labs require students to develop observational evidence for Newton's First and Second Laws. A force probe and motion sensor are used so that students can measure the position, velocity, and acceleration of a dynamics cart and other objects while simultaneously recording the applied forces.

Labs 8 and 9 (One-dimensional Collisions and **Newton's Third Law and Conservation of Momentum):** In these labs students use the same apparatus to observe momentum changes and impulses during one-dimensional collisions.

Students also observe collision forces on a moment-by-moment basis, using force probes mounted on interacting carts, to explore the meaning and validity of Newton's Third Law.

Lab 10 (Two-dimensional Motion—Projectile Motion): This lab allows students to explore the kinematics and dynamics of two-dimensional motions of objects moving under the influence of visible applied forces and invisible gravitational forces.

Labs 11 and 12 (Work and Energy and **Conservations of Energy):** These labs involve the exploration of work and energy, using an MBL motion sensor to display the kinetic energy of a moving cart in real time. The conservation of mechanical energy is explored in two systems: a low-friction cart on an inclined ramp and a mass oscillating at the end of a spring.

ON-LINE TEACHERS' GUIDE

The *Teachers' Guide* for *RealTime Physics Mechanics* is available on-line at **http:/www.wiley.com/college/sokoloff-physics.** This *Guide* focuses on pedagogical (teaching and learning) aspects of using the curriculum, as well as computer-based and other equipment. The *Guide* is offered as an aid to busy physics educators and does not pretend to delineate the "right" way to use the *RealTime Physics Mechanics* curriculum and certainly not the MBL tools. There are many right ways. The *Guide* does, however, explain the educational philosophy that influenced the design of the curriculum and tools and suggests effective teaching methods. Most of the suggestions have come from the college, university, and high school teachers who have participated in field testing of the curriculum.

The *On-line Teachers' Guide* has thirteen sections. Section I presents suggestions regarding computer hardware and software to aid in the implementation of this activity-based MBL curriculum. Sections II through XIII present information about the twelve different labs. Included in each of these is information about the specific equipment and materials needed, tips on how to optimize student learning, answers to questions in the labs, and complete answers to the homework.

EXPERIMENT CONFIGURATION FILES

Experiment configuration files are used to set up the appropriate software features to go with the activities in these labs. You will either need the set of files which is designed for the software package you are using, or you will need to set up the files yourself. At this writing, experiment configuration files for *RealTime Physics Mechanics* are available with Vernier Software and Technology, *Logger Pro* (for Windows and Macintosh), and with PASCO *Data Studio* (for Macintosh and Windows). Appendix A of this module outlines the features of the experiment configuration files for *RealTime Physics Mechanics.* For more information, consult the *On-line Teachers' Guide.*

CONCLUSIONS

RealTime Physics Mechanics has been used in a variety of different educational settings. Many university, college, and high school faculty who have used this curriculum have reported improvements in student understanding of Newton's

Laws. Their comments are supported by our careful analysis of pre- and post-test data using the *Force and Motion Conceptual Evaluation* reported in the literature.[10,11,12] Similar research on the effectiveness of *RealTime Physics Electric Circuits,*[13] *Heat and Thermodynamics* and *Light and Optics* also show dramatic conceptual learning gains in these topic areas. We feel that by combining the outcomes of physics educational research with microcomputer-based tools, the laboratory can be a place where students acquire both a mastery of difficult physics concepts and vital laboratory skills.

ACKNOWLEDGMENTS

RealTime Physics Mechanics could not have been developed without the hardware and software development work of Stephen Beardslee, Lars Travers, Ronald Budworth, and David Vernier. We are indebted to numerous college, university, and high school physics teachers, and especially Curtis Hieggelke (Joliet Junior College), John Garrett (Sheldon High School), and Maxine Willis (Gettysburg High School) for beta testing earlier versions of the laboratories with their students. At the University of Oregon, we especially thank Dean Livelybrooks for supervising the introductory physics laboratory, for providing invaluable feedback, and for writing some of the homework solutions for the *Teachers' Guide.* Frank Womack, Dan DePonte, and all of the introductory physics laboratory teaching assistants provided valuable assistance and input. We also thank the faculty at the University of Oregon (especially Stan Micklavzina), Tufts University, and Dickinson College for their input into *Tools for Scientific Thinking Motion and Force* and *Workshop Physics* on which parts of *RealTime Physics Mechanics* are based, and for assisting with our conceptual learning assessments. Finally, we could not have even started this project if not for our students' active participation in these endeavors.

This work was supported in part by the National Science Foundation under grant number DUE-9455561, *"Activity Based Physics: Curricula, Computer Tools, and Apparatus for Introductory Physics Courses,"* grant number USE-9150589, *"Student Oriented Science,"* grant number DUE-9451287, *"RealTime Physics II: Active University Laboratories Based on Workshop Physics and Tools for Scientific Thinking,"* grant number USE-9153725, *"The Workshop Physics Laboratory Featuring Tools for Scientific Thinking,"* and grant number TPE-8751481, *"Tools for Scientific Thinking: MBL for Teaching Science Teachers,"* and by the Fund for Improvement of Post-secondary Education (FIPSE) of the U.S. Department of Education under grant number G008642149, *"Tools for Scientific Thinking,"* and number P116B90692, *"Interactive Physics."*

REFERENCES

1. Ronald K. Thornton and David R. Sokoloff, "RealTime Physics: Active Learning Laboratory," in *The Changing Role of the Physics Department in Modern Universities, Proceedings of the International Conference on Undergraduate Physics Education,* 1101–1118 (American Institute of Physics, 1997).

2. Ronald K. Thornton and David R. Sokoloff, "Tools for Scientific Thinking—Heat and Temperature Curriculum and Teachers' Guide," (Portland, Vernier Software, 1993) and David R. Sokoloff and Ronald K. Thornton, "Tools for Scientific Thinking—Motion and Force Curriculum and Teachers' Guide," Second edition, (Portland, Vernier Software, 1992).

3. P. W. Laws, "Calculus-based Physics Without Lectures," *Phys. Today* **44**: 12, 24–31 (December, 1991).

4. Priscilla W. Laws, *Workshop Physics Activity Guide: The Core Volume with Module 1: Mechanics,* (Wiley, New York, 1997).

5. David R. Sokoloff and Ronald K. Thornton, "Using Interactive Lecture Demonstrations to Create an Active Learning Environment," *The Physics Teacher* **27**: 6, 340 (1997).

6. L. C. McDermott, "Millikan lecture 1990: What We Teach and What is Learned—Closing the Gap," *Am. J. Phys* **59**, 301 (1991).

7. Milton A. Rothman, Discovering the Natural Laws; the Experimental Basis of Physics, (Doubleday, New York, 1972), Chapter 2.

8. A. B. Arons, *A Guide to Introductory Physics Teaching,* (Wiley, New York, 1990).

9. Priscilla W. Laws, "A New Order for Mechanics," in *Conference on the Introductory Physics Course,* J.W. Wilson, ed. (Wiley, New York, 1997), pp. 125–136.

10. Ronald K. Thornton and David R. Sokoloff, "Assessing Student Learning of Newton's Laws: The *Force and Motion Conceptual Evaluation* and the Evaluation of Active Learning Laboratory and Lecture Curricula," *Am. J. Phys.* **64**, 338 (1998).

11. Ronald K. Thornton, "Learning Physics Concepts in the Introductory Course: Microcomputer-based Labs and Interactive Lecture Demonstrations," in *Conference on the Introductory Physics Course,* J.W. Wilson, ed. (Wiley, New York, 1997), pp. 69–85.

12. Ronald K. Thornton and David R. Sokoloff, "Learning Motion Concepts Using Real-Time Microcomputer-Based Laboratory Tools," *Am. J. Phys.* **58**, 858 (1990).

13. David R. Sokoloff, "Teaching Electric Circuit Concepts Using Microcomputer-Based Current and Voltage Probes," chapter in *Microcomputer-Based Labs: Educational Research and Standards,* Robert F. Tinker, ed., *Series F, Computer and Systems Sciences,* **156**, 129–146 (Berlin, Heidelberg, Springer Verlag, 1996).

This project was supported, in part, by the Fund for the Improvement of Post-Secondary Education (FIPSE) and the National Science Foundation. Opinions expressed are those of the authors and not necessarily those of the foundations.

Contents

Name________________________________ Date__________________

Pre-Lab Preparation Sheet for Lab 1: Introduction To Motion

(Due at the beginning of Lab 1)

Directions:

Read over Lab 1 and then answer the following questions about the procedures.

1. In Activity 1-1, part 3, how do you think graph a will differ from graph b?

2. What can you say in general about velocity versus time for the graphs a, b, and c in Activity 1-3, part 3?

3. Draw your graph for Prediction 2-1 below:

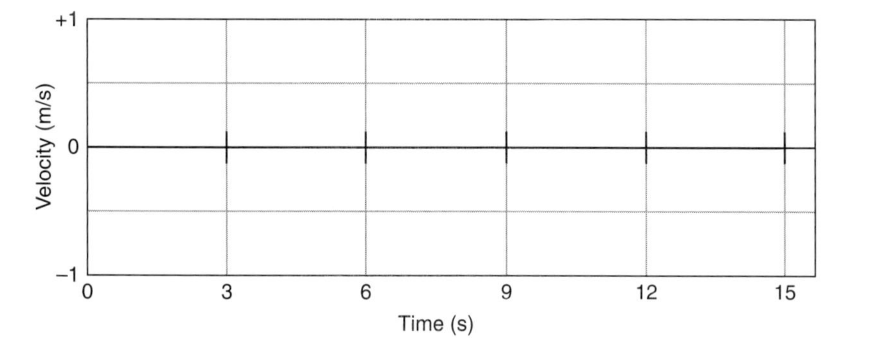

4. In Activity 3-2, how will you find the average velocity?

5. What is a vector? What vector quantities are studied in this lab?

Name_______________ Date_______ Partners_______________

Lab 1: Introduction to Motion

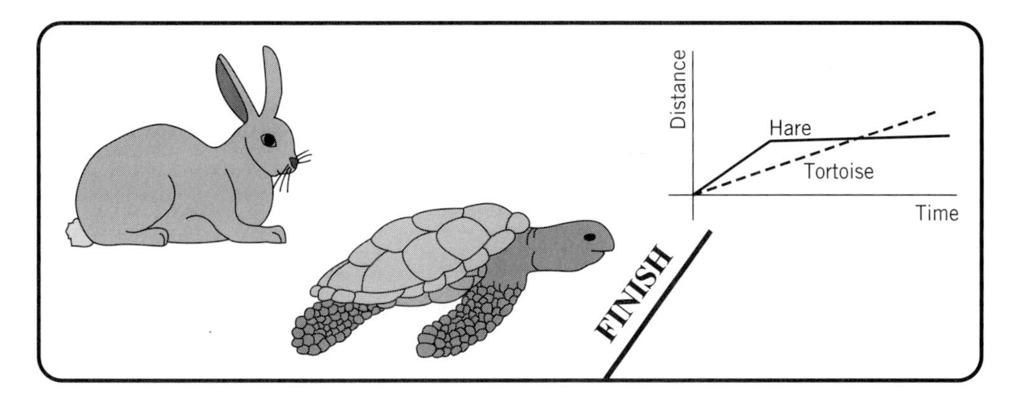

Slow and steady wins the race.

—Aesop's fable: The Hare and the Tortoise

OBJECTIVES

- To discover how to use a motion detector.
- To explore how various motions are represented on a distance (position)–time graph.
- To explore how various motions are represented on a velocity–time graph.
- To discover the relationship between position–time and velocity–time graphs.
- To begin to explore acceleration–time graphs.

OVERVIEW

In this lab you will examine two different ways that the motion of an object that moves along a line can be represented graphically. You will use a motion detector to plot distance–time (position–time) and velocity–time graphs of the motion of your own body and a cart. The study of motion and its mathematical and graphical representation is known as *kinematics*.

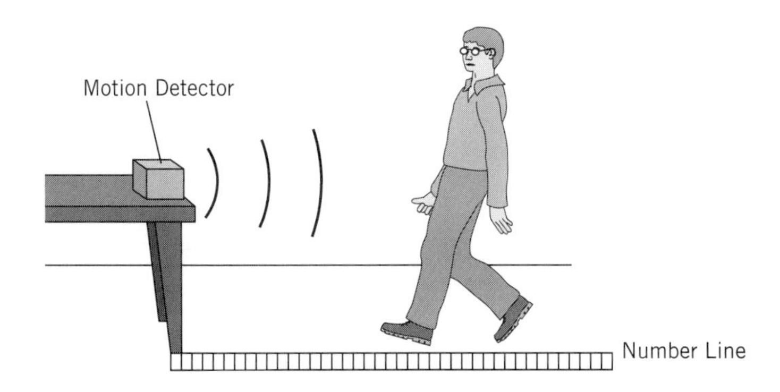

INVESTIGATION 1: DISTANCE (POSITION)–TIME GRAPHS OF YOUR MOTION

The purpose of this investigation is to learn how to relate graphs of the distance as a function of time to the motions they represent.

You will need the following materials:

- computer-based laboratory system
- motion detector
- *RealTime Physics Mechanics* experiment configuration files
- number line on floor in meters (optional)

How does the distance–time graph look when you move slowly? Quickly? What happens when you move toward the motion detector? Away? After completing this investigation, you should be able to look at a distance–time graph and describe the motion of an object. You should also be able to look at the motion of an object and sketch a graph representing that motion.

Comment: "Distance" is short for "distance from the motion detector." The motion detector is the origin from which distances are measured. The motion detector

- detects the closest object directly in front of it (including your arms if you swing them as you walk).
- transfers information to the computer via the interface so that as you walk (or jump, or run), the graph on the computer screen displays your distance from the motion detector.
- will not correctly measure anything closer than some distance (usually specified by the manufacturer).

When making your graphs, don't go closer than this distance from the motion detector.

Activity 1-1: Making and Interpreting Distance–Time Graphs

1. Be sure that the interface is connected to the computer, and the motion detector is plugged into the appropriate port of the interface. Open the experiment file called **Distance (L01A1-1a)** to display distance (position) vs. time axes.
2. If you have a number line on the floor and you want the detector to produce readings that agree, stand at the 2-m mark on the number line, **begin graphing,** and have someone move the detector until the reading is 2 m.
3. **Begin graphing** and make distance–time graphs for different walking speeds and directions, and sketch your graphs on the axes.

a. Start at the 1/2-meter mark and make a distance-time graph, walking away from the detector (origin) *slowly and steadily.*

Distance (m)

Time (s)

b. Make a distance-time graph, walking away from the detector (origin) *medium fast and steadily.*

Distance (m)

Time (s)

c. Make a distance-time graph, walking toward the detector (origin) *slowly and steadily.*

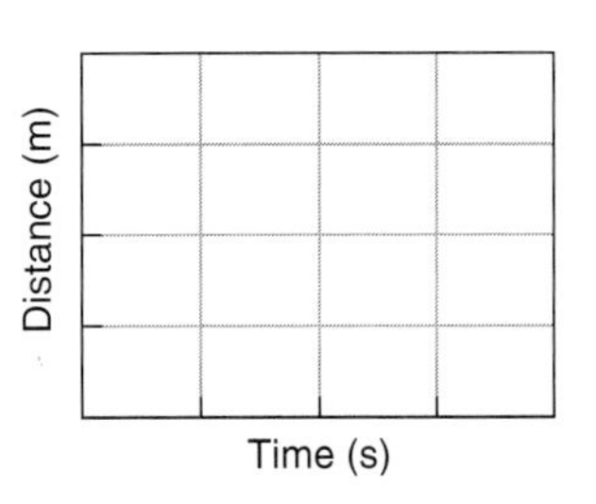

d. Make a distance-time graph, walking toward the detector (origin) *medium fast and steadily.*

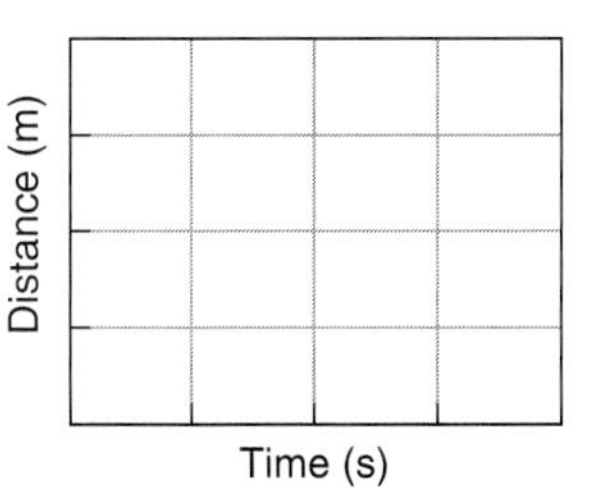

Question 1-1: Describe the difference between a graph made by walking away *slowly* and one made by walking away *quickly*.

Question 1-2: Describe the difference between a graph made by walking *toward* and one made walking *away from* the motion detector.

> **Comment:** It is common to refer to the distance of an object from some origin as the *position* of the object. Since the motion detector is at the origin of the coordinate system, it is better to refer to the graphs you have made as *position–time* graphs rather than distance–time graphs.

Prediction 1-1: Predict the position–time graph produced when a person starts at the 1-m mark, walks away from the detector slowly and steadily for 5 s, stops for 5 s, and then walks toward the detector twice as fast. Draw your prediction on the left axes below using a dashed line.

Compare your predictions with those made by others in your group. Draw your group's prediction on the left-hand axes below using a solid line. (Do not erase your original prediction.)

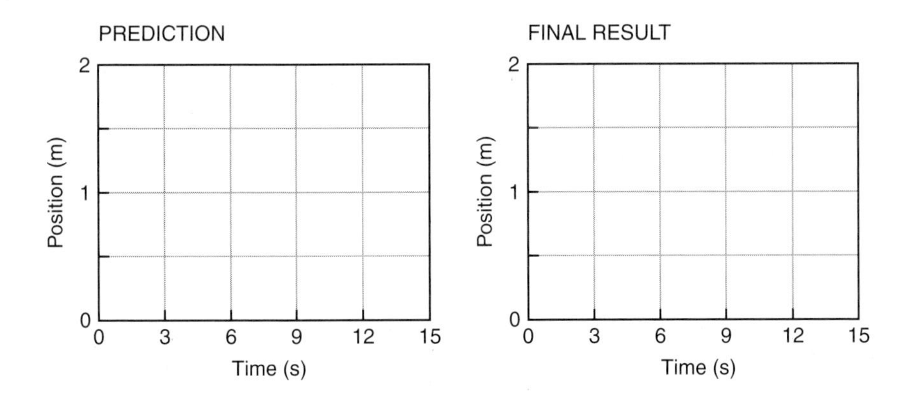

4. Test your prediction. Open the experiment file called **Away and Back (L01A1-1b)** to set up the software to graph position over a range of 2 m for a time interval of 15 s.

   Move in the way described in Prediction 1-1, and graph your motion. When you are satisfied with your graph, draw your group's final result on the right axes above.

Question 1-3: Is your prediction the same as the final result? If not, describe how you would move to make a graph that looks like your *prediction.*

Activity 1-2: Matching a Position–Time Graph

By now you should be pretty good at predicting the shape of a position–time graph of your movements. Can you do things the other way around by reading a position–time graph and figuring out how to move to reproduce it? In this activity you will move to match a position graph shown on the computer screen.

1. Open the experiment file called **Position Match (L01A1-2).** A position graph like that shown below will appear on the screen. **Clear** any other data remaining from previous experiments.

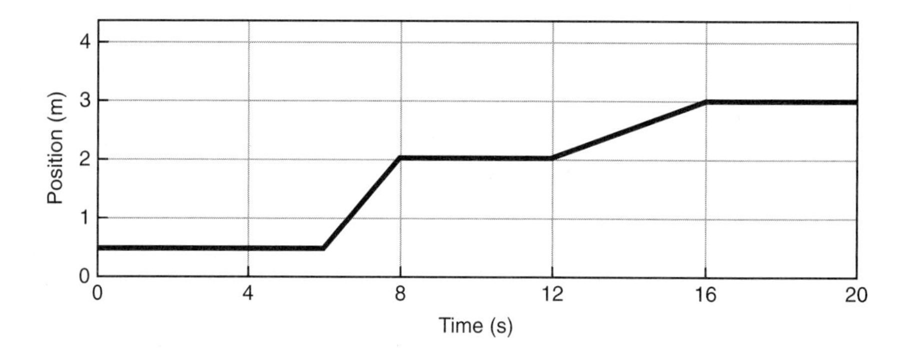

> **Comment:** This graph is stored in the computer so that it is **persistently displayed on the screen.** New data from the motion detector can be collected without erasing the **Position Match** graph.

2. Move to match the **Position Match** graph on the computer screen. You may try a number of times. It helps to work in a team. Get the times right. Get the positions right. Each person should take a turn.

Question 1-4: What was the difference in the way you moved to produce the two differently sloped parts of the graph you just matched?

Activity 1-3: Other Position–Time Graphs

Note: **Clear** the **Position Match** graph from the screen before moving on.

1. Sketch your own position–time graph on the axes which follow with a dashed line. Use straight lines, no curves. Now see how well someone in your group can duplicate this graph on the screen by walking in front of the motion detector.

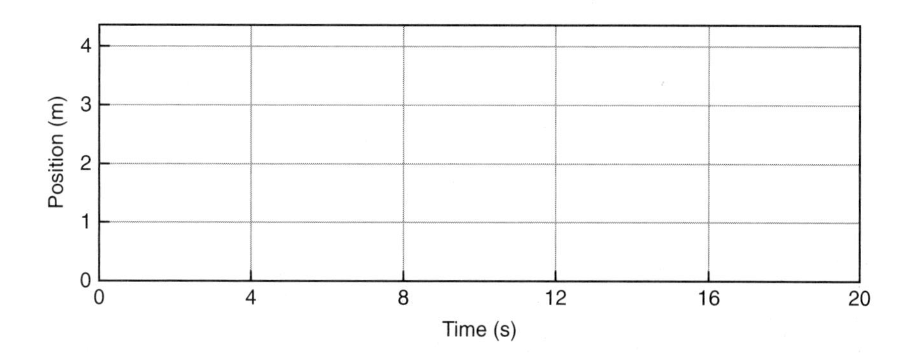

2. Draw the best attempt by a group member to match your position–time graph on the same axes. Use a solid line.

3. Can you make a curved position–time graph? Try to make each of the graphs shown below.

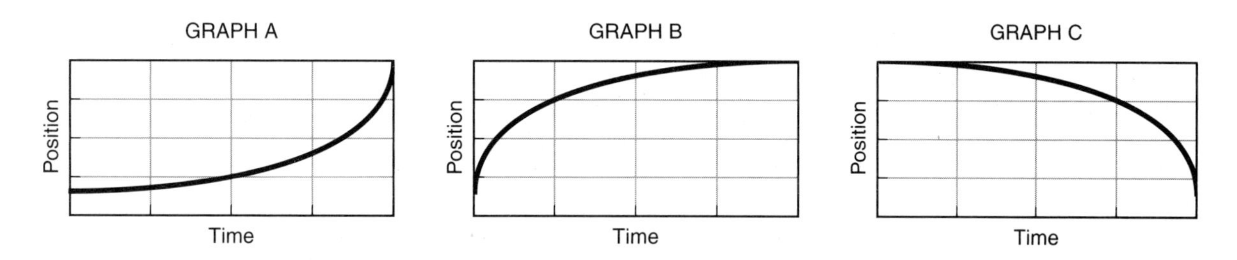

4. Describe how you must move to produce a position–time graph with each of the shapes shown.

   Graph A answer:

   Graph B answer:

Graph C answer:

Question 1-5: What is the general difference between motions that result in a *straight-line* position–time graph and those that result in a *curved-line* position–time graph?

INVESTIGATION 2: VELOCITY–TIME GRAPHS OF MOTION

You have already plotted your position along a line as a function of time. Another way to represent your motion during an interval of time is with a graph that describes how fast and in what direction you are moving. This is a *velocity–time* graph. *Velocity* is the rate of change of position with respect to time. It is a quantity that takes into account your speed (how fast you are moving) and also the direction you are moving. Thus, when you examine the motion of an object moving along a line, the direction the object is moving is indicated by the sign (positive or negative) of the velocity.

Graphs of velocity over time are more challenging to create and interpret than those for position. A good way to learn to interpret them is to create and examine velocity–time graphs of your own body motions, as you will do in this investigation.

You will need the following materials:

- computer-based laboratory system
- motion detector
- *RealTime Physics Mechanics* experiment configuration files
- number line on floor in meters (optional)

Activity 2-1: Making Velocity Graphs

1. Set up to graph velocity. Open the experiment file called **Velocity Graphs (L01A2-1)** to set up the axes that follow.
2. Graph your velocity for different walking speeds and directions as described in (a)–(d) below, and sketch your graphs on the axes. *(Just draw smooth patterns; leave out smaller bumps that are mostly due to your steps.)*
   a. **Begin graphing** and make a velocity graph by walking *away* from the detector *slowly and steadily*. Try again until you get a graph you're satisfied with.

      You may want to **adjust the velocity scale** so that the graph fills more of the screen and is clearer.

      Then sketch your graph on the axes.

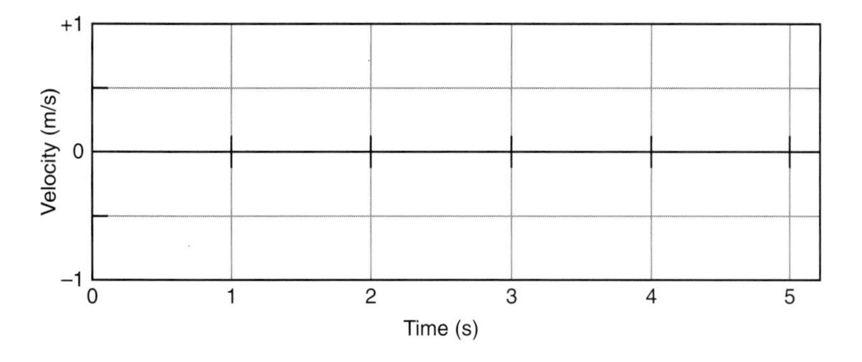

b. Make a velocity graph, walking *away* from the detector *medium fast and steadily.*

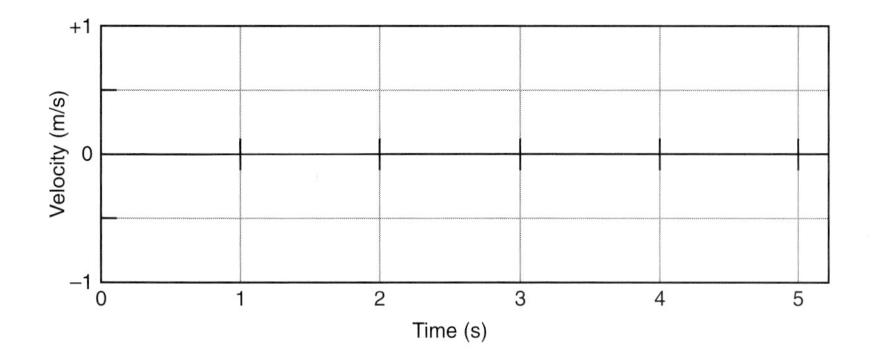

c. Make a velocity graph, walking *toward* the detector *slowly and steadily.*

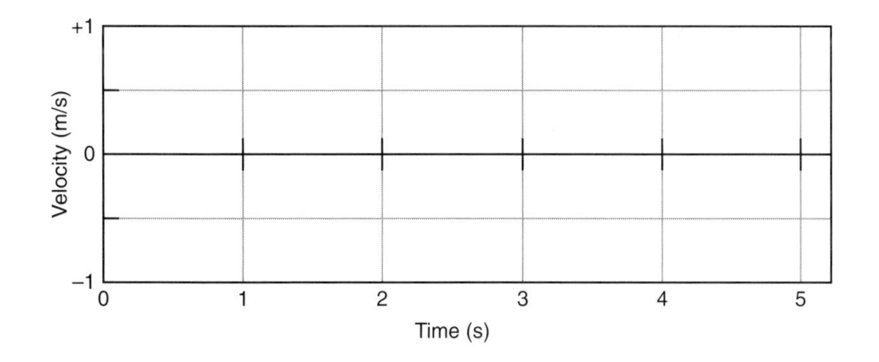

d. Make a velocity graph, walking *toward* the detector *medium fast and steadily.*

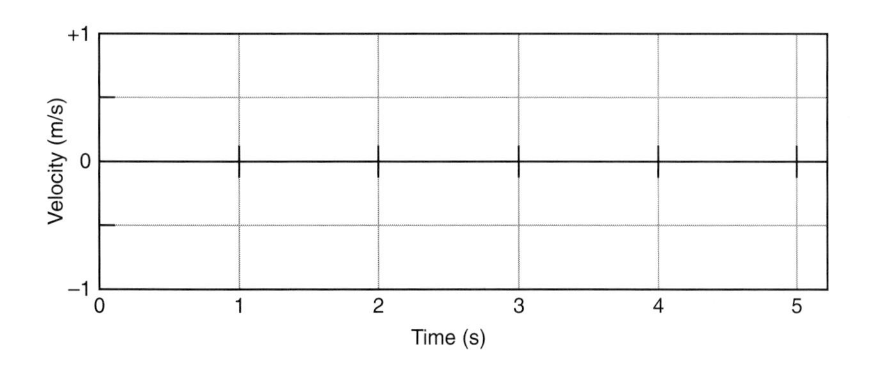

Question 2-1: What is the most important difference between the graph made by *slowly* walking away from the detector and the one made by walking away *more quickly?*

Question 2-2: How are the velocity–time graphs different for motion *away* and motion *toward* the detector?

Prediction 2-1: Predict a velocity–time graph for a more complicated motion and check your prediction.

Each person draw below, using a *dashed line,* your *prediction* of the velocity–time graph produced if you

- walk away from the detector slowly and steadily for about 5 s;
- stand still for about 5 s;
- walk toward the detector steadily about twice as fast as before.

Compare your predictions and see if you can all agree. Use a solid line to draw in your group prediction.

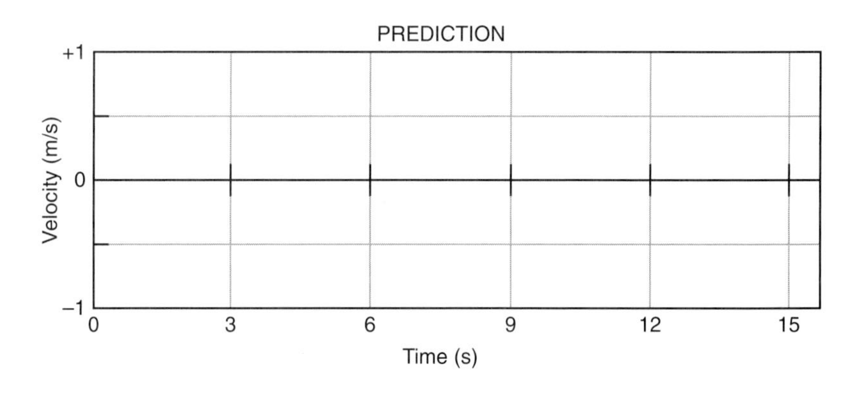

3. Test your prediction. (Be sure to **adjust the time scale** to 15 s.) **Begin graphing** and repeat your motion until you think it matches the description.

   Draw the best graph on the axes below. *Be sure the 5 s you spend standing still shows clearly.*

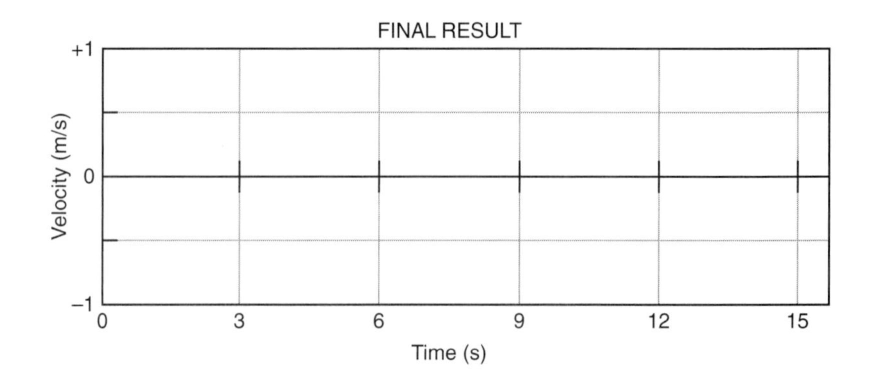

Comment: Velocity implies both speed and *direction.* How fast you move is your speed: the rate of change of position with respect to time. As you have seen, for motion along a line (e.g., the positive x axis) the sign (+ or −) of the velocity indicates the direction. If you move away from the detector (origin), your velocity is positive, and if you move toward the detector, your velocity is negative.

The faster you move *away* from the origin, the larger positive number your velocity is. The faster you move *toward* the origin, the "larger" negative number your velocity is. That is −4 m/s is twice as fast as −2 m/s, and both motions are toward the origin.

These two ideas of speed and direction can be combined and represented by *vectors.* A velocity vector is represented by an arrow pointing in the direction of motion. The length of the arrow is drawn proportional to the speed; the longer the arrow, the larger the speed. If you are moving toward the right, your velocity vector can be represented by

If you were moving twice as fast toward the right, the arrow representing your velocity vector would look like

while moving twice as fast toward the left would be represented by

What is the relationship between a one-dimensional velocity vector and the *sign* of velocity? This depends on the way you choose to set the positive x axis.

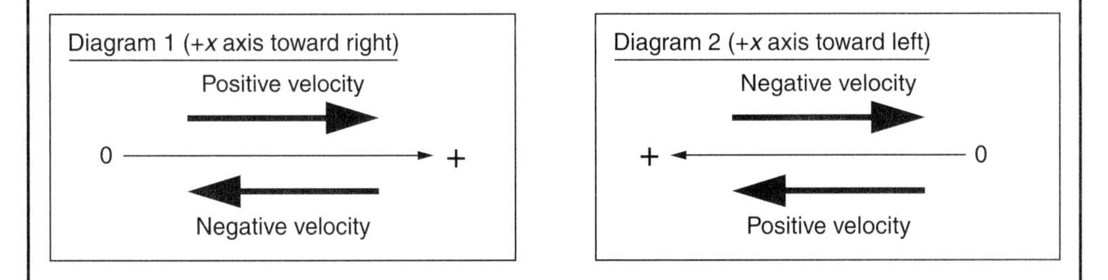

In both diagrams, the top vectors represent velocity toward the right. In Diagram 1, the x axis has been drawn so that the positive x direction is toward the right, as it is usually drawn. Thus, the top arrow represents *positive* velocity. However, in Diagram 2, the positive x direction is toward the left. Thus the top arrow represents *negative* velocity. Likewise, in both diagrams the bottom arrows represent velocity toward the left. In Diagram 1 this is *negative* velocity, and in Diagram 2 it is *positive* velocity.

Question 2-3: Sketch below velocity vectors representing the three parts of the motion described in Prediction 2-1.

Walking slowly away from the detector:

Standing still:

Walking rapidly toward the detector:

Activity 2-2: Matching a Velocity Graph

In this activity, you will try to move to match a velocity–time graph shown on the computer screen. This is often much harder than matching a position graph as you did in the previous investigation. Most people find it quite a challenge at first to move so as to match a velocity graph. In fact, some velocity graphs that can be invented cannot be matched!

1. Open the experiment file called **Velocity Match (L01A2-2)** to display the velocity–time graph shown below on the screen.

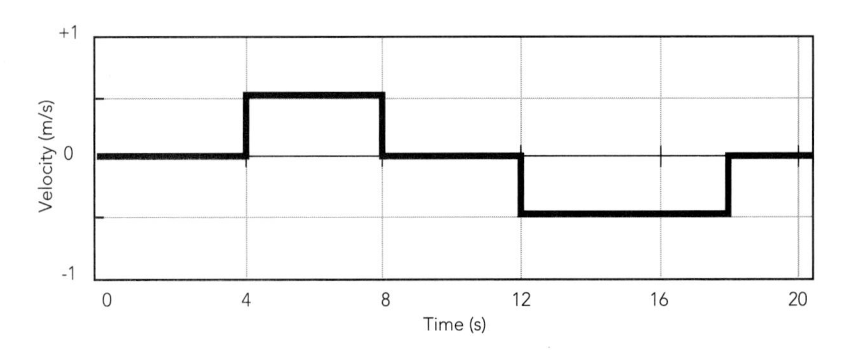

Prediction 2-2: Describe in words how you would move so that your velocity matched each part of this velocity–time graph.

0 to 4 s:

4 to 8 s:

8 to 12 s:

12 to 18 s:

18 to 20 s:

2. **Begin graphing,** and move so as to imitate this graph. You may try a number of times. Work as a team and plan your movements. Get the times right. Get the velocities right. Each person should take a turn.

   Draw in your group's best match on the axes above.

Question 2-4: Describe how you moved to match each part of the graph. Did this agree with your predictions?

Question 2-5: Is it possible for an object to move so that it produces an absolutely vertical line on a velocity–time graph? Explain.

Question 2-6: Did you run into the motion detector on your return trip? If so, why did this happen? How did you solve the problem? Does a velocity graph tell you where to start? Explain.

If you have more time, do the following Extension.

Extension 2-3: More Velocity Graphs

Prediction E2-3: Can you tell from a velocity–time graph where you were when you started walking and where you were when you stopped walking? Explain.

Set up the motion detector and test your prediction. Make a graph walking away from or toward the motion detector. Use the features of your software to transfer your data so that the graph will remain **persistently displayed on the screen.** Then, for comparison, make a second graph moving in exactly the same manner, but this time start at a different distance from the detector. Sketch the graphs. Compare them.

Question E2-7: If someone showed you these graphs, could you tell from the graphs where the person was when s/he started moving? Explain.

Prediction E2-4: You are driving down the highway, and another car moving in the same direction passes you. At the moment that the car passes you, which motion quantity is the same for both cars—position, velocity, or both?

Devise a method to test your prediction. Set up the motion detector to look at two people walking in the same direction next to one another at the moment one passes the other. (**Hint:** Remember that the motion detector detects the object that is closest to it.) Graph the motions. **Print** your graphs. Compare the results to your prediction.

Question E2-8: What is your conclusion? Which quantities are the same when one object passes another—their velocities, positions, or both?

INVESTIGATION 3: RELATING POSITION AND VELOCITY GRAPHS

You have looked at position–time and velocity–time graphs separately. Since position–time and velocity–time graphs are different ways to represent the same motion, it is possible to figure out the velocity at which someone is moving by examining her/his position–time graph. Conversely, you can also figure out how far someone has traveled (change in position) from a velocity–time graph.

To explore how position–time and velocity–time graphs are related, you will need the following materials:

- computer-based laboratory system
- motion detector
- *RealTime Physics Mechanics* experiment configuration files
- number line on floor in meters (optional)

Activity 3-1: Predicting Velocity Graphs From Position Graphs

1. Open the experiment file called **Velocity from Position (L01A3-1)** to set up the axes shown that follow. **Clear** any previous graphs.

Prediction 3-1: Predict a velocity graph from a position graph. Carefully study the position–time graph that follows and predict the velocity–time graph that would result from the motion. Using a *dashed line,* sketch your *prediction* of the corresponding velocity–time graph on the velocity axes.

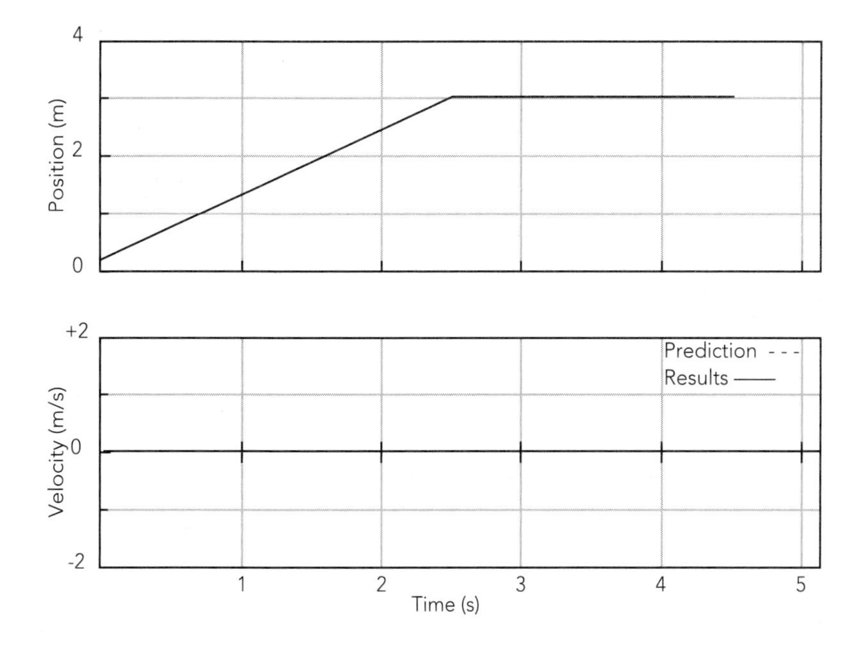

2. Test your prediction. After each person has sketched a prediction, **begin graphing,** and do your group's best to make a position graph like the one shown. Walk as smoothly as possible.

   When you have made a good duplicate of the position graph, sketch your actual graph over the existing position–time graph.

   Use a *solid line* to draw the actual velocity–time graph on the same axes with your prediction. (Do not erase your prediction.)

Question 3-1: How would the position graph be different if you moved faster? Slower?

Question 3-2: How would the velocity graph be different if you moved faster? Slower?

Activity 3-2: Calculating Average Velocity

In this activity, you will find an average velocity from your velocity–time graph in Activity 3-1 and then from your position–time graph.

1. Find your average velocity from your velocity graph in Activity 3-1. Use the **analysis feature** in the software to read values of velocity (about 10 values *from the portion of your velocity graph where your velocity is relatively constant*) and use them to calculate the average (mean) velocity. Write the 10 values in the table that follows.

Velocity values (m/s)			
1		6	
2		7	
3		8	
4		9	
5		10	

Average (mean) value of the velocity: ______m/s

Comment: Average velocity during a particular time interval can also be calculated as the change in position divided by the change in time. (The change in position is often called the *displacement*.) For motion with a constant velocity, this is also the *slope* of the position–time graph for that time period.

As you have observed, the faster you move, the steeper your position–time graph becomes. The *slope* of a position–time graph is a quantitative measure of this incline. The size of this number tells you the speed, and the sign tells you the direction.

2. Calculate your average velocity from the slope of your position graph in Activity 3-1. Use the **analysis feature** of the software to read the position and time coordinates for two typical points *while you were moving*. (For a more accurate answer, use two points as far apart as possible but still typical of the motion, and within the time interval in which you took velocity readings in part 1.)

	Position (m)	Time (s)
Point 1		
Point 2		

Calculate the change in position (displacement) between points 1 and 2. Also calculate the corresponding change in time *(time interval)*. Divide the change in position by the change in time to calculate the *average* velocity. Show your calculations below.

Change in position (m)	
Time interval (s)	
Average velocity (m/s)	

Question 3-3: Is the average velocity positive or negative? Is this what you expected?

Question 3-4: Does the average velocity you just calculated from the position graph agree with the average velocity you found from the velocity graph? Do you expect them to agree? How would you account for any differences?

If you have additional time, do the Extension below, and use other features of the software to find the average velocity from your graphs. If you are planning to come back to this later, be sure to **save** your graphs now.

Extension 3-3: Using Statistics and Fit to Find the Average Velocity

In Activity 3-2, you found the value of the average velocity for a steady motion in two ways: from the average of a number of values on a velocity–time graph and from the slope of the position–time graph. The **statistics feature** in your software should allow you to find the average (mean) value directly from the velocity–time graph. The **fit routine** should allow you to find the line that best fits your position–time graph from Activity 3-1. The equation of this line includes a value for the slope.

1. *Using* ***statistics:*** You must first **select the portion** of the velocity–time graph for which you want to find the mean value. Next use the **statistics feature** to read the mean value of velocity during this portion of the motion.

Question E3-5: Compare this value to the one you found from 10 measurements in Activity 3-2. Which method do you think is more accurate? Why?

2. *Using* ***fit:*** You must first **select the portion** of the position–time graph that you want to fit.

   Next, use the **fit routine,** select a linear fit, $y = a + bt$, and then find the equation of the line.

   Record the equation of the fit line below, and compare the value of the slope (b) to the velocity you found in Activity 3-2.

Question E3-6: What is the meaning of a?

Question E3-7: How do the two values of velocity that you found here agree with each other? Is this what you expected?

Activity 3-4: Predicting Position Graphs From Velocity Graphs

Prediction 3-2: Carefully study the velocity graph shown below. Using a *dashed line,* sketch your *prediction* of the corresponding position graph on the bottom set of axes. (Assume that you started at the 1-m mark.)

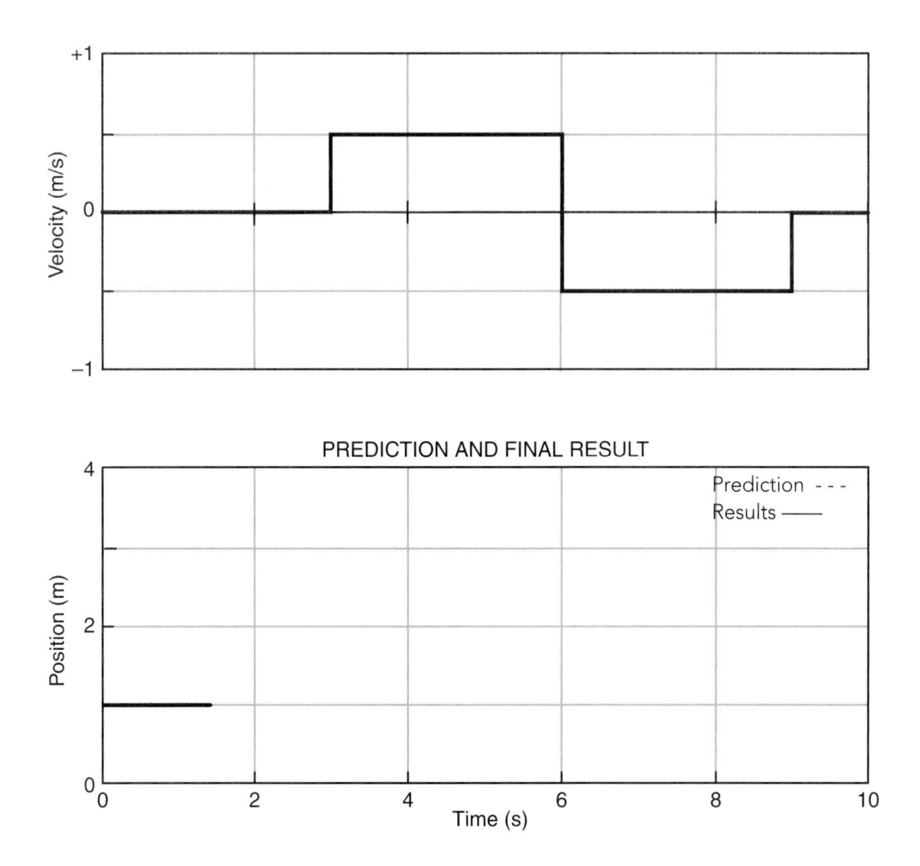

1. Test your prediction. First shut off the **analysis feature,** and **adjust the time axis** to 0 to 10 s before you start.

2. After each person has sketched a prediction, do your group's best to duplicate the top (velocity–time) graph by walking. Be sure to **graph velocity first.**

   When you have made a good duplicate of the velocity–time graph, draw your actual result over the existing velocity–time graph.

3. Use a *solid line* to draw the actual position–time graph on the same axes with your prediction. (Do not erase your prediction.)

Question 3-8: How can you tell from a *velocity*–time graph that the moving object has changed direction? What is the velocity at the moment the direction changes?

Question 3-9: How can you tell from a position–time graph that your motion is steady (motion at a constant velocity)?

Question 3-10: How can you tell from a velocity–time graph that your motion is steady (constant velocity)?

INVESTIGATION 4: INTRODUCTION TO ACCELERATION

There is a third quantity besides position and velocity that is used to describe the motion of an object—acceleration. Acceleration is defined as the *rate of change of velocity with respect to time* (just like velocity is defined as the *rate of change of position with respect to time*). In this investigation you will begin to examine the acceleration of objects.

Because of the jerky nature of the motion of your body, the acceleration graphs are very complex. It will be easier to examine the motion of a cart. In this investigation you will examine the cart moving with a constant (steady) velocity. Later, in Lab 2 you will examine the acceleration of more complex motions of the cart. You will need the following:

- computer-based laboratory system
- motion detector
- *RealTime Physics Mechanics* experiment configuration files
- cart with very little friction
- smooth ramp or other level surface 2–3 m long

Activity 4-1: Motion of a Cart at a Constant Velocity

To graph the motion of a cart at a constant velocity you can give the cart a quick push with your hand and then release it.

1. Set up the motion detector at the end of the ramp. If the cart has a friction pad, move it out of contact with the ramp so that the cart can move freely.

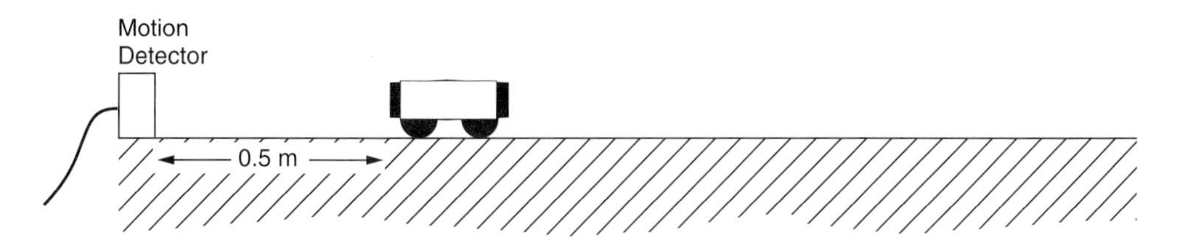

2. Set up the position and velocity axes that follow by opening the experiment file called **Constant Velocity (L01A4-1).**

Prediction 4-1: How should the position and velocity graphs look if you move the cart at a constant velocity away from the motion detector starting at the 0.5-m mark? Sketch your predictions with dashed lines on the axes that follow. **Hint:** base your prediction on your observations of the motion of your body in Investigations 1 and 2.

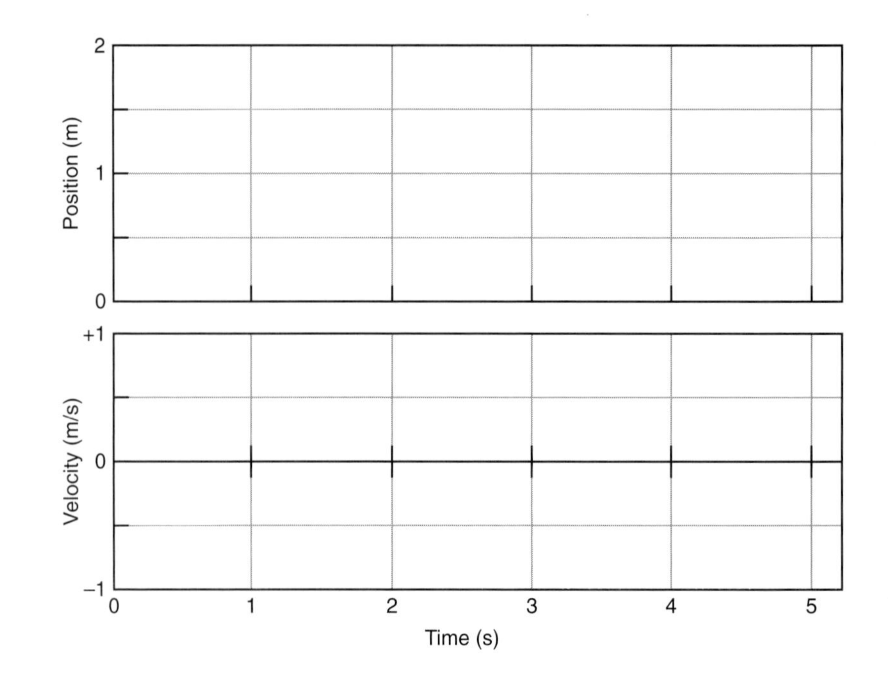

3. Test your prediction. *Be sure that the cart is never closer than 0.5 m from the motion detector and that your hand is not between the cart and motion detector.* **Begin graphing.** Try several times until you get a fairly constant velocity. Sketch your results with solid lines on the axes.

Question 4-1: Did your position–time and velocity–time graphs agree with your predictions? What characterizes constant velocity motion on a position–time graph?

Question 4-2: What characterizes constant velocity motion on a velocity–time graph?

Activity 4-2: Acceleration of a Cart Moving at a Constant Velocity

Prediction 4-2: Sketch with a dashed line on the axes that follow your prediction of the acceleration of the cart you just observed moving at a constant velocity away from the motion detector. Base your prediction on the definition of acceleration.

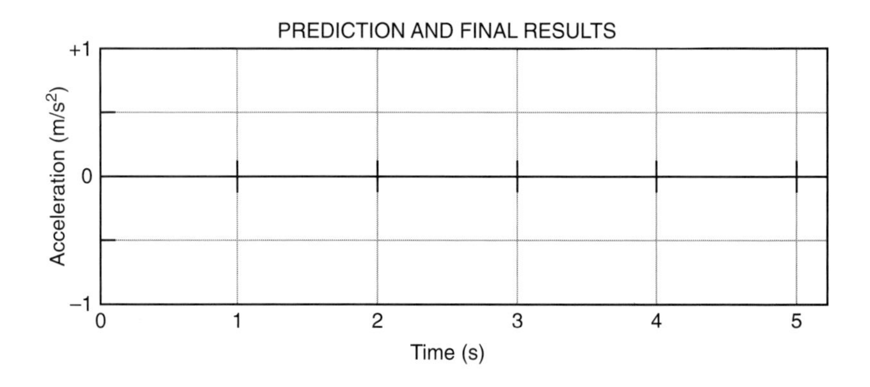

4. **Display** the real acceleration graph of the cart in place of the position graph. Adjust the axes as necessary to display acceleration clearly. Sketch the acceleration graph using a solid line on the axes above.

> **Comment:** To find the average acceleration of the cart during some time interval (the average rate of change of its velocity with respect to time), you must measure its velocity at the beginning and end of the interval, calculate the difference between the final value and the initial value and divide by the time interval.

Question 4-3: Does the acceleration–time graph you observed agree with this method of calculating acceleration? Explain. Does it agree with your prediction? What is the value of the acceleration of an object moving at a constant velocity?

Question 4-4: The diagram below shows positions of the cart at equal time intervals. (This is like overlaying snapshots of the cart at equal time intervals. The motion detector also looks at the cart's position at equal intervals.) At each indicated time, sketch a vector above the cart that might represent the velocity of the cart at that time while it is moving at a constant velocity away from the motion detector. Assume that the cart is already moving at t_1.

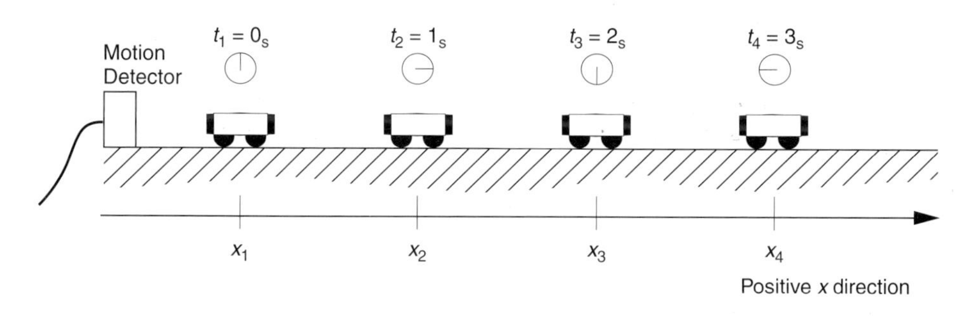

> **Comment:** To find the average acceleration vector from two velocity vectors, you must first find the vector representing the *change in velocity* by subtracting the initial velocity vector from the final one. Then you divide this vector by the time interval.

Question 4-5: Show below how you would find the vector representing the change in velocity between the times 2 and 3 s in the diagram in Question 4-4. (**Hint:** The vector difference is the same as the sum of one vector and the negative of the other vector.) From this vector, what value would you calculate for the acceleration? Explain. Is this value in agreement with the acceleration graph on the previous page?

Name______________________ Date____________ Partners______________________

Homework for Lab 1: Introduction to Motion

POSITION–TIME GRAPHS

Answer the following questions in the spaces provided.

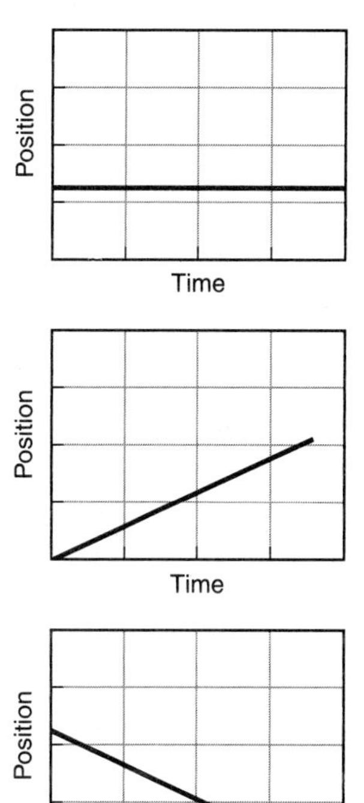

1. What do you do to create a horizontal line on a position–time graph?

2. How do you walk to create a straight line that slopes up?

3. How do you walk to create a straight line that slopes down?

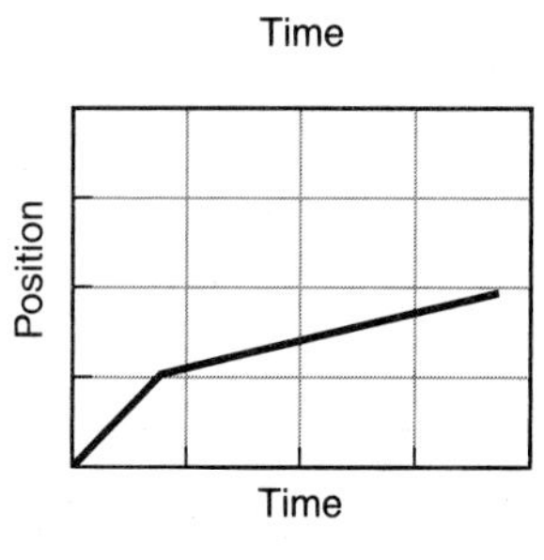

4. How do you move so the graph goes up steeply at first, and then continues up less steeply?

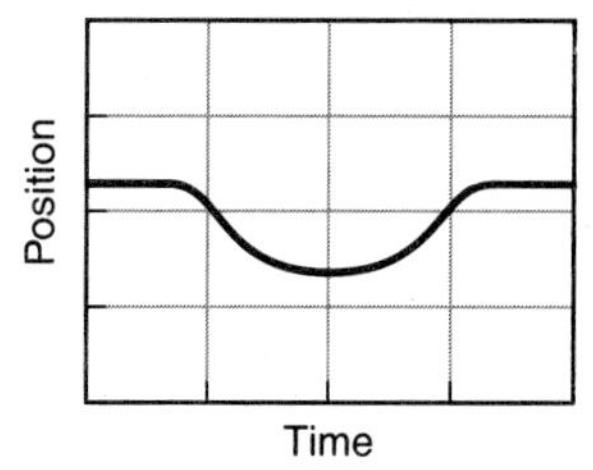

5. How do you walk to create a U-shaped graph?

Answer the following about two objects, A and B, whose motion produced the following position–time graphs.

6. **a.** Which object is moving faster—A or B?

   b. Which starts ahead? Define what you mean by "ahead."

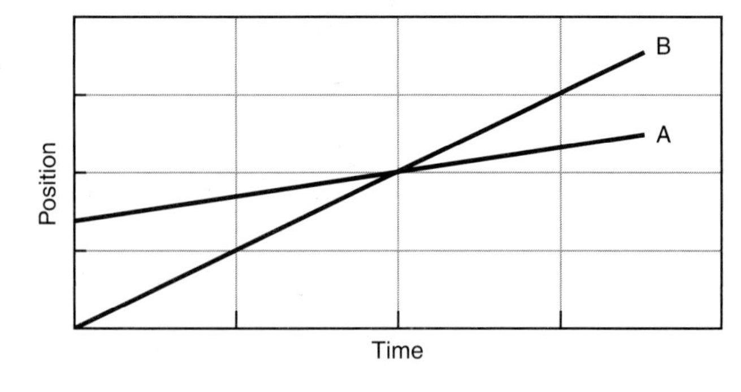

   c. What does the intersection mean?

7. **a.** Which object is moving faster?

   b. Which object has a negative velocity according to the convention we have established?

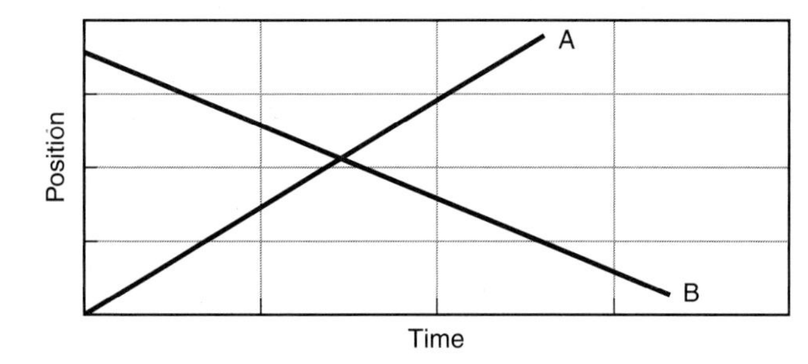

8. **a.** Which object is moving faster?

   b. Which starts ahead? Explain what you mean by "ahead."

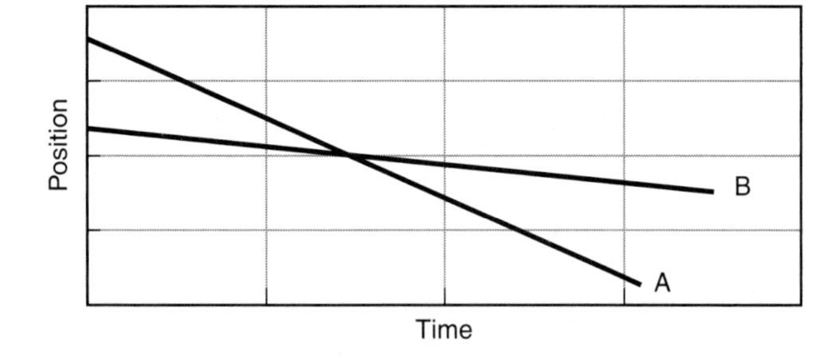

Sketch the position–time graph corresponding to each of the following descriptions of the motion of an object.

9. The object moves with a steady (constant) velocity away from the origin.

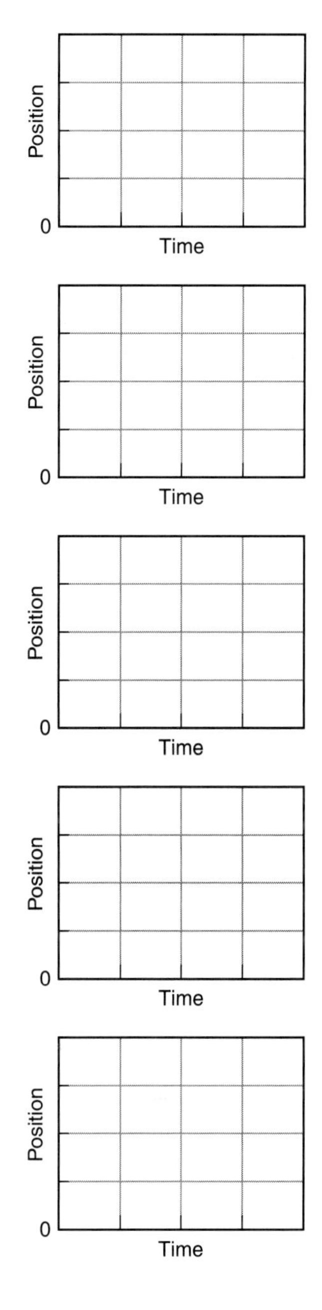

10. The object is standing still.

11. The object moves with a steady (constant) velocity toward the origin for 5 s and then stands still for 5 s.

12. The object moves with a steady velocity away from the origin for 5 s, then reverses direction and moves at the same speed toward the origin for 5 s.

13. The object moves away from the origin, starting slowly and speeding up.

VELOCITY–TIME GRAPHS

After studying the velocity–time graphs you have made, answer the following questions.

1. How do you move to create a horizontal line in the positive part of a velocity–time graph, as shown on the right?

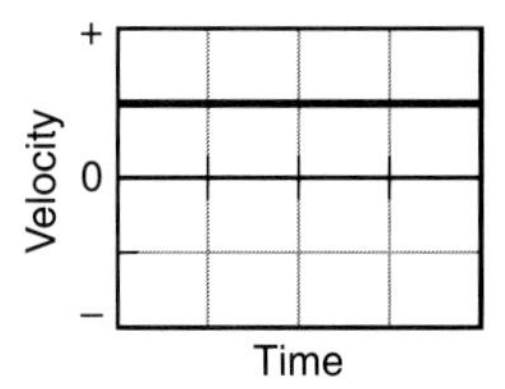

2. How do you move to create a straight-line velocity–time graph that slopes up from zero, as shown on the right?

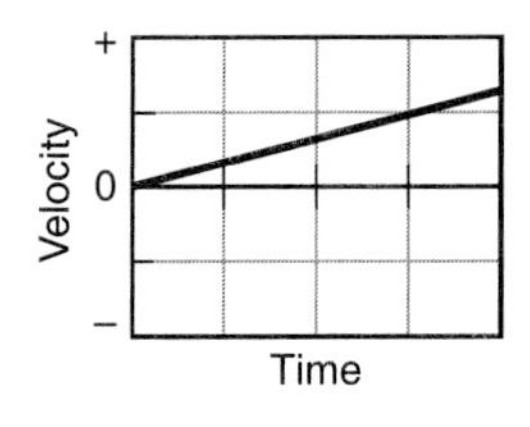

3. How do you move to create a straight-line velocity–time graph that slopes down, as shown on the right?

4. How do you move to make a horizontal line in the negative part of a velocity–time graph, as shown on the right?

5. The velocity–time graph of an object is shown below. Figure out the total change in position *(displacement)* of the object. Show your work.

   Displacement = ______m

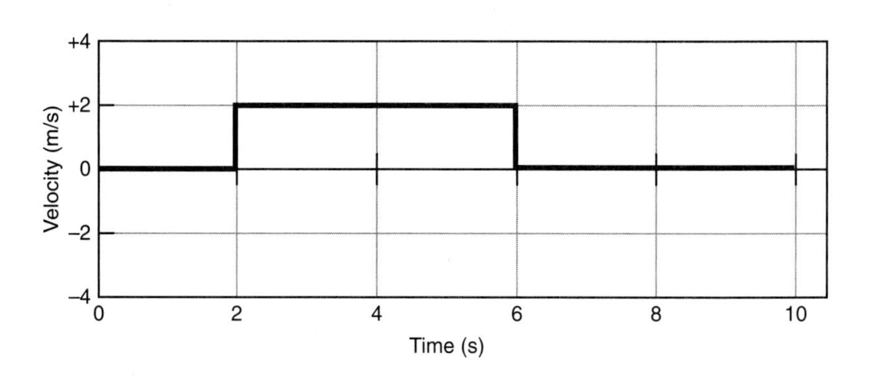

6. Both of the *velocity* graphs below show the motion of two objects, A and B. Answer the following questions separately for 1 and for 2. Explain your answers when necessary.

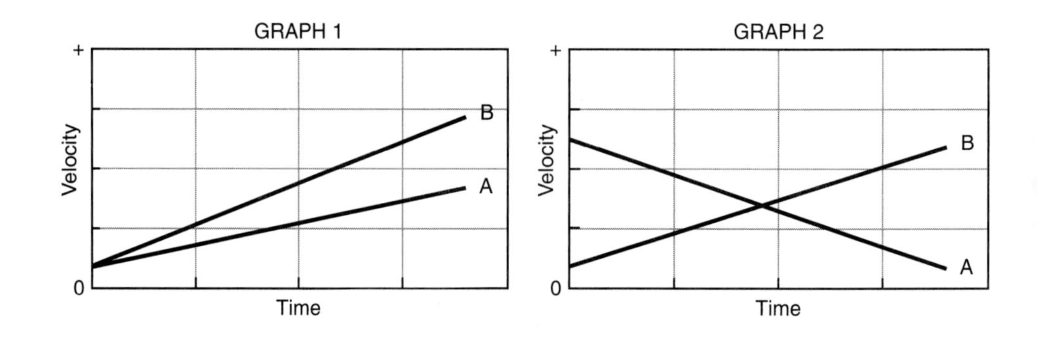

a. Is one faster than the other? If so, which one is faster—A or B?

b. What does the intersection mean?

c. Can you tell which object is "ahead"? (Define "ahead.")

d. Does either A or B reverse direction? Explain.

a. Is one faster than the other? If so, which one is faster—A or B?

b. What does the intersection mean?

c. Can you tell which object is "ahead"? (Define "ahead.")

d. Does either A or B reverse direction? Explain.

Sketch the velocity–time graph corresponding to each of the following descriptions of the motion of an object.

7. The object is moving away from the origin at a steady (constant) velocity.

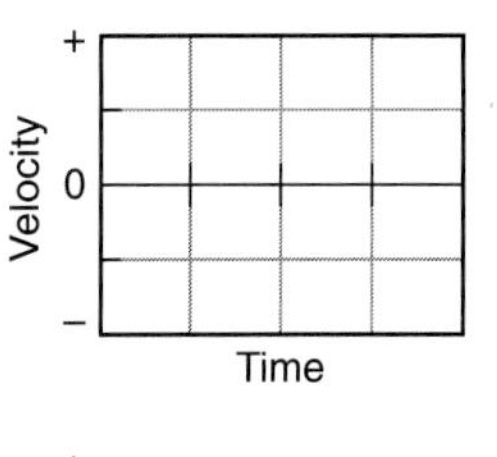

8. The object is standing still.

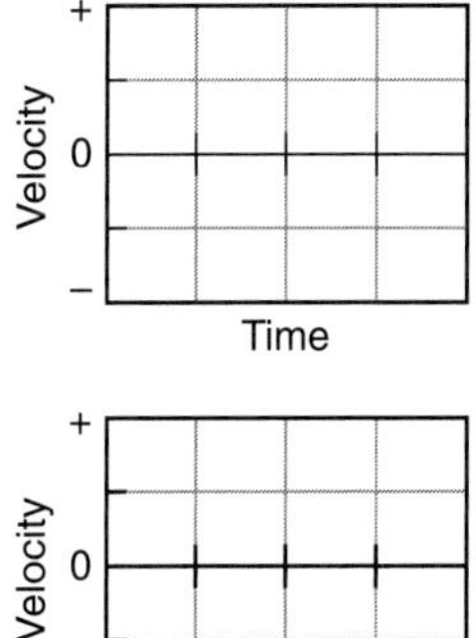

9. The object moves toward the origin at a steady (constant) velocity for 10 s, and then stands still for 10 s.

+
Velocity
0
–
0 10 20
Time (s)

10. The object moves away from the origin at a steady (constant) velocity for 10 s, reverses direction, and moves back toward the origin at the same speed for 10 s.

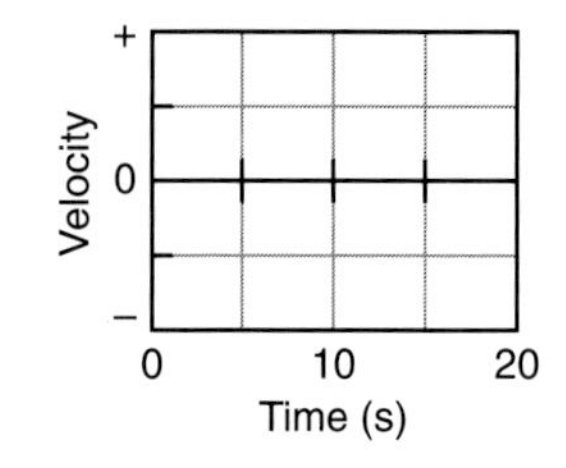

11. Draw the velocity graphs for an object whose motion produced the position–time graphs shown below on the left. Position is in meters (m) and velocity in meters per second (m/s). (**Note:** Unlike most real objects, you can assume that these objects can change velocity so quickly that it looks instantaneous with this time scale.)

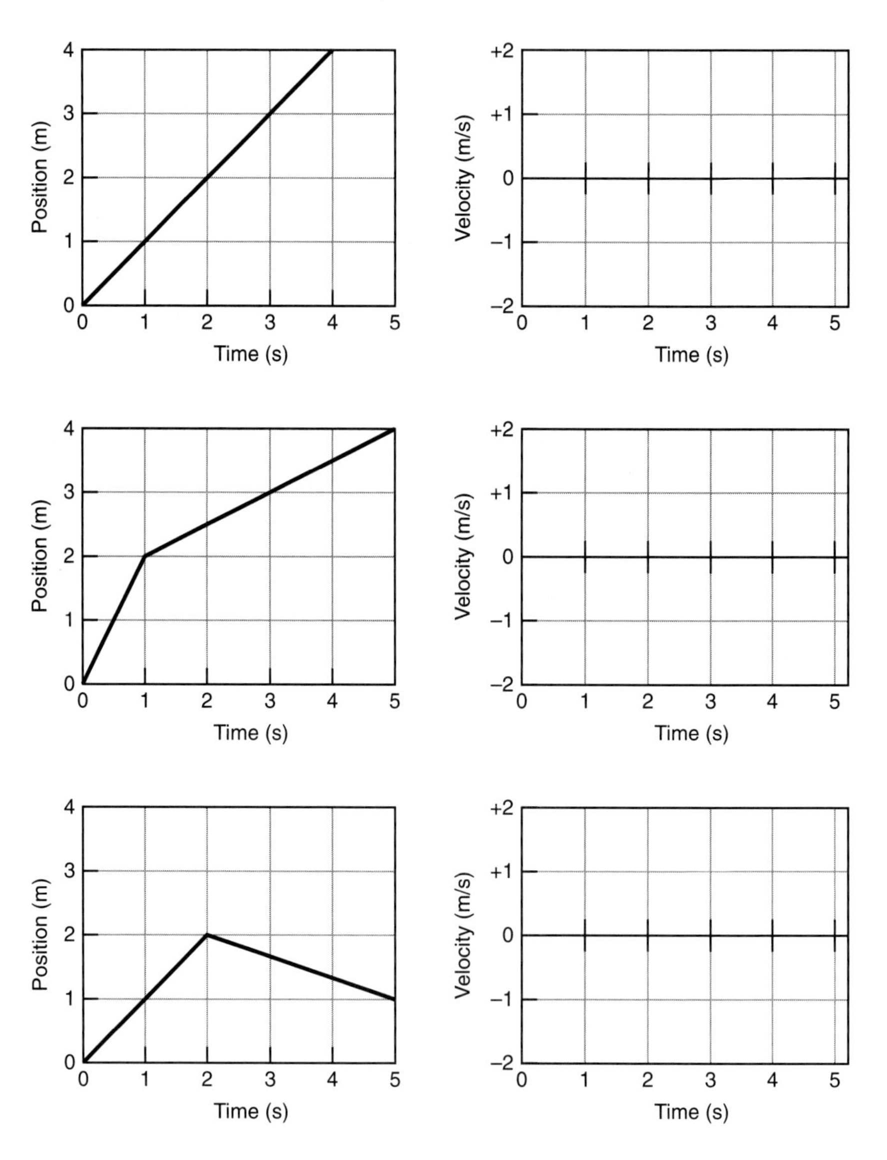

12. Draw careful graphs below of position and velocity for a cart that

a. moves away from the origin at a slow and *steady* (constant) velocity for the first 5 s.

b. moves away at a medium-fast, *steady* (constant) velocity for the next 5 s.

c. stands still for the next 5 s.

d. moves toward the origin at a slow and *steady* (constant) velocity for the next 5 s.

e. stands still for the last 5 s.

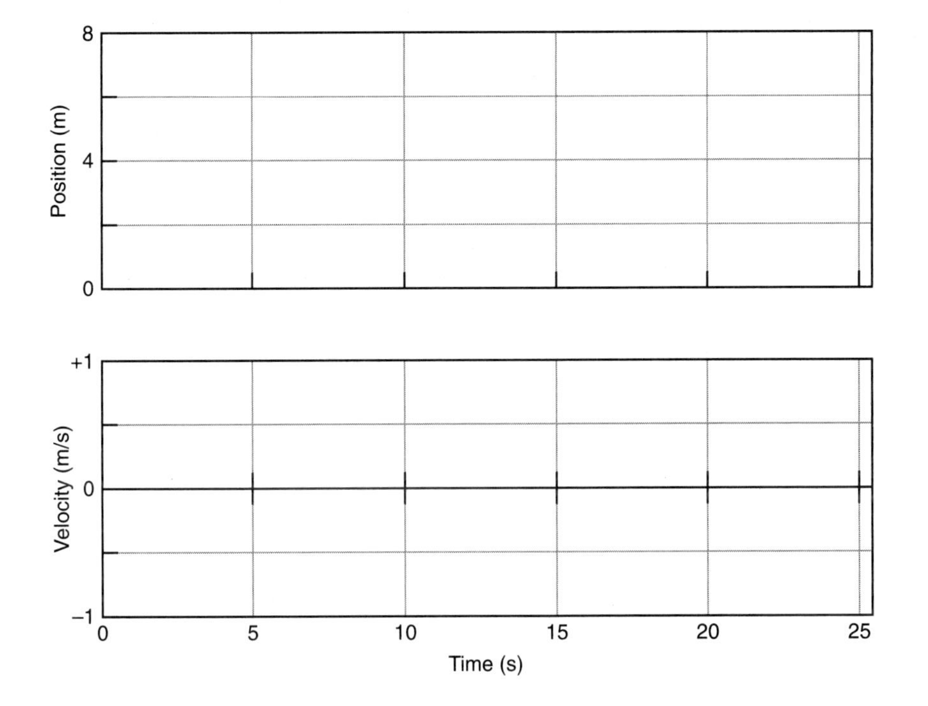

Name______________________________ Date____________________

Pre-Lab Preparation Sheet for Lab 2: Changing Motion

(Due at the beginning of Lab 2)

Directions:
Read over Lab 2 and then answer the following questions about the procedures.

1. In Activity 1-1, how do you expect that your position–time graph will differ from those you observed in Lab 1, where you were moving with a constant velocity?

2. Show how you would add the two vectors shown below:

$$\rightarrow + \longrightarrow$$

3. Show below how you would subtract the second vector from the first.

$$\rightarrow - \longrightarrow$$

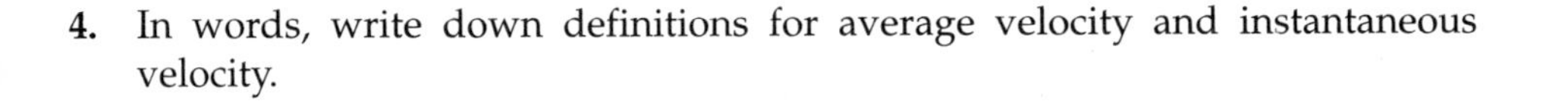

4. In words, write down definitions for average velocity and instantaneous velocity.

5. What do you predict the direction (sign) of acceleration will be for the experiment in Activity 3-2: Speeding Up Toward the Motion Detector?

Name________________________ Date____________ Partners__________________________

Lab 2: Changing Motion

A cheetah can accelerate from 0 to 50 miles per hour in 6.4 seconds.

—Encyclopedia of the Animal World

A Jaguar can accelerate from 0 to 50 miles per hour in 6.1 seconds.

—World Cars

OBJECTIVES

- To discover how and when objects accelerate.
- To understand the meaning of acceleration, its magnitude, and its direction.
- To discover the relationship between velocity and acceleration graphs.
- To learn how to represent velocity and acceleration using vectors.
- To learn how to find average acceleration from acceleration graphs.
- To learn how to calculate average acceleration from velocity graphs.

OVERVIEW

In the previous lab, you looked at position–time and velocity–time graphs of the motion of your body and a cart at a constant velocity. You also looked at the acceleration–time graph of the cart. The data for the graphs were collected using a motion detector. Your goal in this lab is to learn how to describe various kinds of motion in more detail.

You have probably realized that a velocity–time graph is easier to use than a position–time graph when you want to know how fast and in what direction you are moving at each instant in time as you walk (even though you can calculate this information from a position–time graph).

It is not enough when studying motion in physics to simply say that "the object is moving toward the right" or "it is standing still." When the velocity of an object is changing, it is also important to describe how it is changing. The rate of change of velocity with respect to time is known as the *acceleration.*

To get a feeling for acceleration, it is helpful to create and learn to interpret velocity–time and acceleration–time graphs for some relatively simple motions of a cart on a smooth ramp or other level surface. You will be observing the cart with the motion detector as it moves with its velocity changing at a constant rate.

INVESTIGATION 1: VELOCITY AND ACCELERATION GRAPHS

In this investigation you will be asked to predict and observe the shapes of velocity–time and acceleration–time graphs of a cart moving along a smooth ramp or other level surface. You will focus on cart motions with a steadily increasing velocity.

You will need the following materials:

- computer-based laboratory system
- motion detector
- *RealTime Physics Mechanics* experiment configuration files
- cart with very little friction
- smooth ramp or other level surface 2–3 m long
- fan unit attachment with batteries and dummy cells (or with a speed adjustment control)

Activity 1-1: Speeding Up

In this activity you will look at velocity–time and acceleration–time graphs of the motion of a cart, and you will be able to see how these two representations of the motion are related to each other when the cart is speeding up.

This could be done by moving the cart with your hand, but it is difficult to get a smoothly changing velocity in this way. Instead you will use a fan or propeller driven by an electric motor to accelerate the cart.

1. Set up the cart on the ramp, with the fan unit and motion detector as shown below. Tape the fan unit securely to the cart. *Be sure that the ramp is level. Be sure that the fan blade does not extend beyond the end of the cart facing the motion detector. (If it does, the motion detector may collect bad data from the rotating blade.)*
2. If the cart has a friction pad, move it out of contact with the ramp so that the cart can move freely.

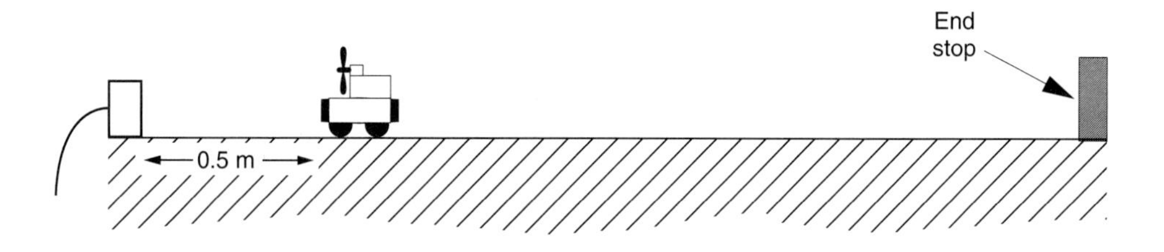

3. Open the experiment file called **Speeding Up (L02A1-1)** to display the axes that follow.
4. Use a position graph to make sure that the detector can "see" the cart all the way to the end of the ramp. You may need to tilt the detector up slightly.
5. Make sure the switch is off, then place half batteries and half dummy cells in the battery compartment of the fan unit (or use all batteries, and set the dial

at about half maximum speed of the fan blade). *To preserve the batteries, switch on the fan unit only when you are making measurements.*

6. Hold the cart with your hand on its side, **begin graphing,** switch the fan unit on and when you hear the clicks of the motion detector, release the cart from rest. *Do not put your hand between the cart and the detector. Be sure to stop the cart before it hits the end stop. Turn off the fan unit.*

   Repeat, if necessary, until you get a nice set of graphs.

   Adjust the position and velocity axes if necessary so that the graphs fill the axes. Use the features of your software to transfer your data so that the graphs will remain **persistently displayed on the screen.**

   Also **save your data** for analysis in Investigation 2. (Name your file **SPEEDUP1.XXX,** where **XXX** are your initials.)

7. Sketch your position and velocity graphs neatly on the axes that follow. Label the graphs *"Speeding Up 1."* (Ignore the acceleration axes for now.)

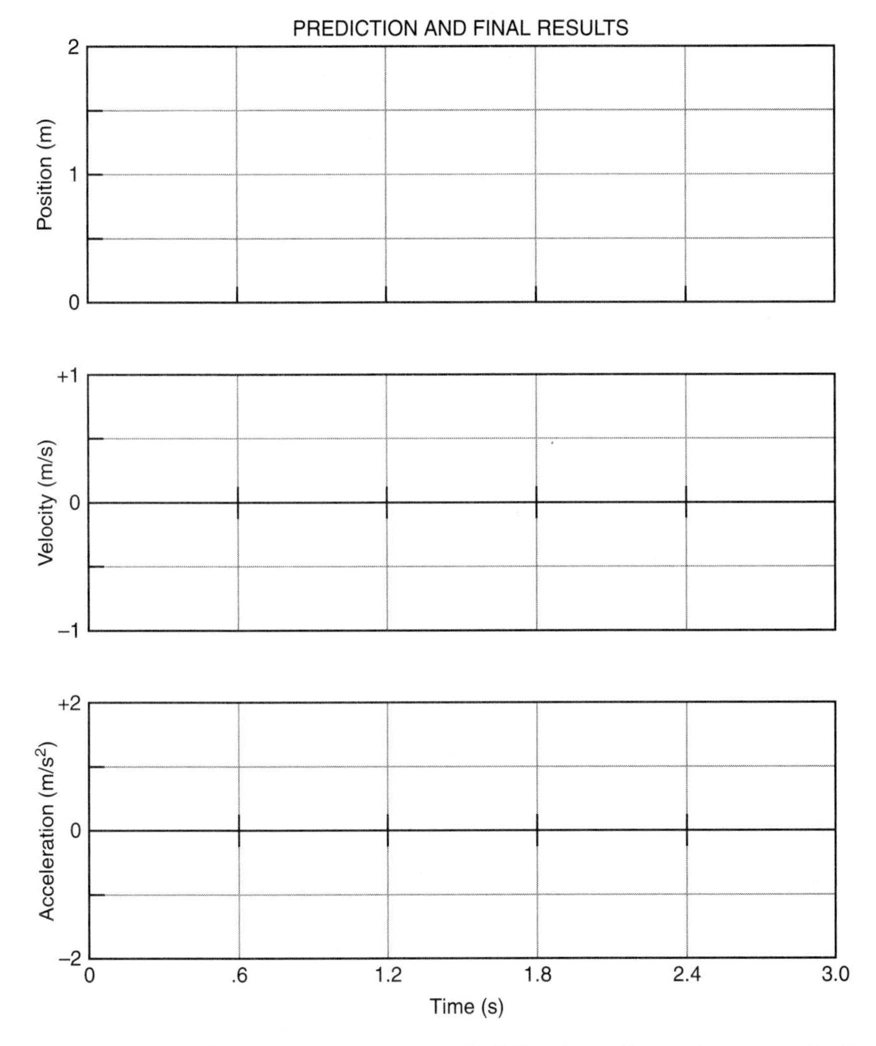

Question 1-1: How does your position graph differ from the position graphs for steady (constant velocity) motion that you observed in Lab 1: Introduction to Motion?

Question 1-2: What feature of your velocity graph signifies that the motion was *away* from the motion detector?

Question 1-3: What feature of your velocity graph signifies that the cart was *speeding up?* How would a graph of motion with a constant velocity differ?

8. Adjust the acceleration scale so that your graph fills the axes. Sketch your graph on the acceleration axes above, and label it *"Speeding Up 1."*

Question 1-4: During the time that the cart is speeding up, is the acceleration positive or negative? How does *speeding up* while moving *away* from the detector result in this sign of acceleration? (**Hint:** Remember that acceleration is the *rate of change* of velocity. Look at how the velocity is changing. It takes two points on the velocity–time graph to calculate the rate of change of velocity.)

Question 1-5: How does the velocity vary in time as the cart speeds up? Does it increase at a steady (constant) rate or in some other way?

Question 1-6: How does the acceleration vary in time as the cart speeds up? Is this what you expect based on the velocity graph? Explain.

Question 1-7: The diagram below shows the positions of a cart at equal time intervals as it speeds up. Assume that the cart is already moving at t_1.

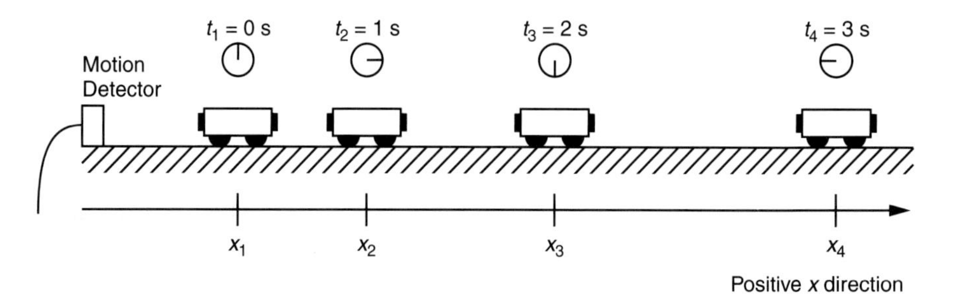

At each indicated time, sketch a vector above the cart that might represent the velocity of the cart at that time while it is moving away from the motion detector and speeding up.

Question 1-8: Show below how you would find the vector representing the change in velocity between the times 2 and 3 s in the diagram above. (**Hint:** Remember that the change in velocity is the final velocity minus the initial velocity, and the vector difference is the same as the sum of one vector and the negative of the other vector.)

Based on the direction of this vector and the direction of the positive x axis, what is the sign of the acceleration? Does this agree with your answer to Question 1-4?

Activity 1-2: Speeding Up More

Prediction 1-1: Suppose that you accelerate the cart at a faster rate. How would your velocity and acceleration graphs be different? Sketch your predictions with dashed or different color lines on the previous set of axes.

1. Test your predictions. Make velocity and acceleration graphs. This time accelerate the cart with the maximum number of batteries in the battery compartment (or set the dial to the maximum speed of the fan blade). *Remember to switch the fan unit on only when making measurements.*

   Repeat if necessary to get nice graphs. (Leave the original graphs **persistently displayed on the screen.**) When you get a nice set of graphs, **save your data** as **SPEEDUP2.XXX** for analysis in Investigation 2.

2. Sketch your velocity and acceleration graphs with solid or different color lines on the previous set of axes, or **print** the graphs and affix them over the axes. Be sure that the graphs are labeled *"Speeding Up 1"* and *"Speeding Up 2."*

Question 1-9: Did the shapes of your velocity and acceleration graphs agree with your predictions? How is the magnitude (size) of acceleration represented on a velocity–time graph?

Question 1-10: How is the magnitude (size) of acceleration represented on an acceleration–time graph?

INVESTIGATION 2: MEASURING ACCELERATION

In this investigation you will examine the motion of a cart accelerated along a level surface by a battery driven fan more quantitatively. This analysis will be quantitative in the sense that your results will consist of numbers. You will determine the cart's acceleration from your velocity–time graph and compare it to the acceleration read from the acceleration–time graph.

You will need motion software and the data files you saved from Investigation 1.

Activity 2-1: Velocity and Acceleration of a Cart That Is Speeding Up

1. The data for the cart accelerated along the ramp with half batteries and half dummy cells (Investigation 1, Activity 1-1) should still be **persistently on the screen.** (If not, **load the data** from the file **SPEEDUP1.XXX.**)

   Display velocity and acceleration, and **adjust the axes** if necessary.

2. Sketch the velocity and acceleration graphs again below, or **print,** and affix a copy of the graphs. Correct the scales if necessary.

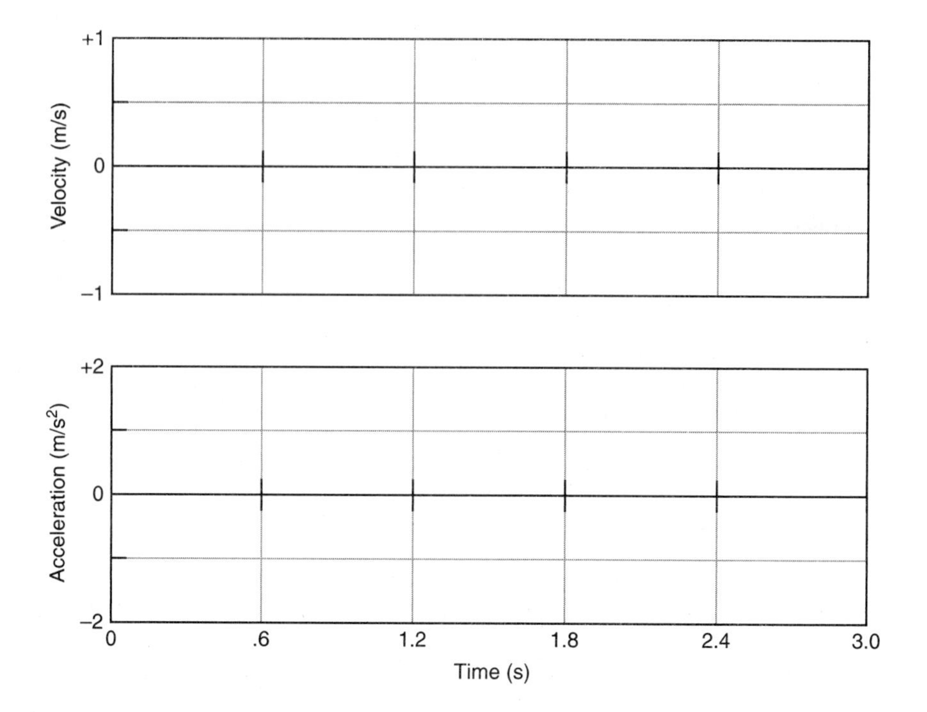

3. Find the average acceleration of the cart from your acceleration graph. Use the **analysis feature** in the software to read a number of values (say 10) of the acceleration, which are equally spaced in time. (Only use values from the portion of the graph after the cart was released and before the cart was stopped.)

Acceleration values (m/s^2)			
1		6	
2		7	
3		8	
4		9	
5		10	

Average (mean) acceleration: _______ m/s^2

Comment: Average acceleration during a particular time interval is defined as the average rate of change of velocity with respect to time—the change in velocity divided by the change in time. By definition, the rate of change of a quantity graphed with respect to time is also the *slope* of the curve. Thus, the (average) slope of an object's velocity–time graph is also the (average) acceleration of the object.

4. Calculate the slope of your velocity graph. Use the **analysis feature** of your software to read the velocity and time coordinates for two typical points on the velocity graph. (For a more accurate answer, use two points as far apart in time as possible but still during the time the cart was speeding up.)

	Velocity (m/s)	Time (s)
Point 1		
Point 2		

Calculate the change in velocity between points 1 and 2. Also calculate the corresponding change in time (time interval). Divide the change in velocity by the change in time. This is the *average* acceleration. Show your calculations below.

Speeding up	
Change in velocity (m/s)	
Time interval (s)	
Average acceleration (m/s^2)	

Question 2-1: Is the acceleration positive or negative? Is this what you expected?

Question 2-2: Does the average acceleration you just calculated agree with the average acceleration you found from the acceleration graph? Do you expect them to agree? How would you account for any differences?

Activity 2-2: Speeding Up More

1. **Load the data** from your file **SPEEDUP2.XXX** (Investigation 1, Activity 1-2). **Display velocity and acceleration.**
2. Sketch the velocity and acceleration graphs or **print** and affix the graphs. Use dashed lines on the previous set of axes.
3. Use the **analysis feature** of the software to read acceleration values, and find the average acceleration of the cart from your acceleration graph.

Acceleration values (m/s^2)			
1		6	
2		7	
3		8	
4		9	
5		10	

Average (mean) acceleration: ______m/s^2

4. Calculate the average acceleration from your velocity graph. Remember to use two points as far apart in time as possible, but still having typical values.

	Velocity (m/s)	Time (s)
Point 1		
Point 2		

Calculate the *average* acceleration.

Speeding up more	
Change in velocity (m/s)	
Time interval (s)	
Average acceleration (m/s^2)	

Question 2-3: Does the average acceleration calculated from velocities and times agree with the average acceleration you found from the acceleration graph? How would you account for any differences?

Question 2-4: Compare this average acceleration to that with half batteries and half dummy cells (Activity 2-1). Which is larger? Is this what you expected?

If you have additional time, do the following Extension.

Extension 2-3: Using Statistics and Fit to Find the Average Acceleration

In Activity 2-1 and 2-2, you found the value of the average acceleration for a motion with steadily increasing velocity in two ways: from the average of a number of values on an acceleration–time graph and from the slope of the velocity–time graph. The **statistics feature** in the software allows you to find the average (mean) value directly from the acceleration–time graph. The **fit routine** allows you to find the line that best fits your velocity–time graph from Activity 2-1 and 2-2. The equation of this line includes a value for the slope.

1. *Using **Statistics:*** **Load** your **SPEEDUP1.XXX** file. You must first **select the portion** of the acceleration–time graph for which you want to find the mean value.

   Next, use the **statistics feature** and read the mean value of acceleration from the table: ______m/s^2

Question E2-5: Compare this value to the one you found from 10 measurements in Activity 2-1.

2. *Using **Fit:*** You must first **select the portion** of the velocity–time graph that you want to fit.

   Next, use the **fit routine** to try a linear fit, $v = b + ct$.

Record the equation of the fit line, and compare the value of the slope *(c)* to the acceleration you found in Activity 2-1.

Question E2-6: What is the physical meaning of *b*?

Question E2-7: How do the two values of acceleration that you found in this extension agree with each other? Is this what you expected?

Find the average acceleration for the motion in your **SPEEDUP2.XXX** file from the acceleration–time and velocity–time graphs using the same methods. Compare the values to those found in Activity 2-2 by averaging 10 values.

INVESTIGATION 3: SLOWING DOWN AND SPEEDING UP

In this investigation you will look at a cart moving along a ramp or other level surface *and slowing down*. A car being driven down a road and brought to rest when the brakes are applied is a good example of this type of motion.

Later you will examine the motion of the cart *toward* the motion detector and *speeding up*.

In both cases, we are interested in how velocity and acceleration change over time. That is, we are interested in the shapes of the velocity–time and acceleration–time graphs (and their relationship to each other), as well as the vectors representing velocity and acceleration.

You will need the following materials:

- computer-based laboratory system
- motion detector
- *RealTime Physics Mechanics* experiment configuration files
- cart with very little friction
- smooth ramp or other level surface 2–3 m long
- fan unit attachment with batteries

Activity 3-1: Slowing Down

In this activity you will look at the velocity and acceleration graphs of the cart moving *away from* the motion detector and *slowing down.*

1. The cart, ramp, and motion detector should be set up as in Investigation 1. Use the maximum number of batteries (or set the dial to the maximum speed). The fan should be pushing the cart *toward* the motion detector.

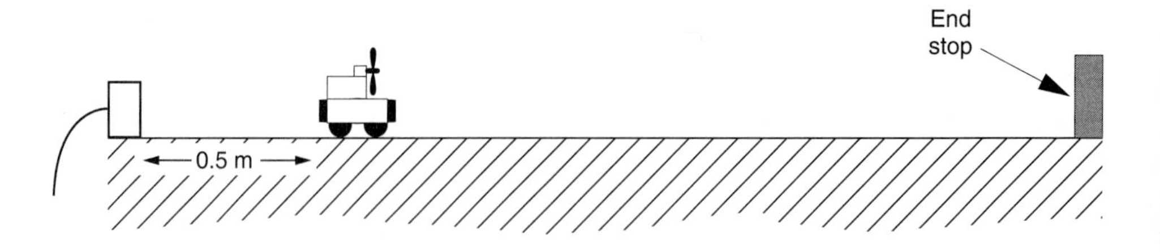

Now, when you give the cart a quick push away from the motion detector with the fan running, it will slow down after it is released.

Prediction 3-1: If you give the cart a short push away from the motion detector and release it, will the acceleration be positive, negative, or zero (after it is released)?

Sketch your predictions for the velocity–time and acceleration–time graphs on the axes below.

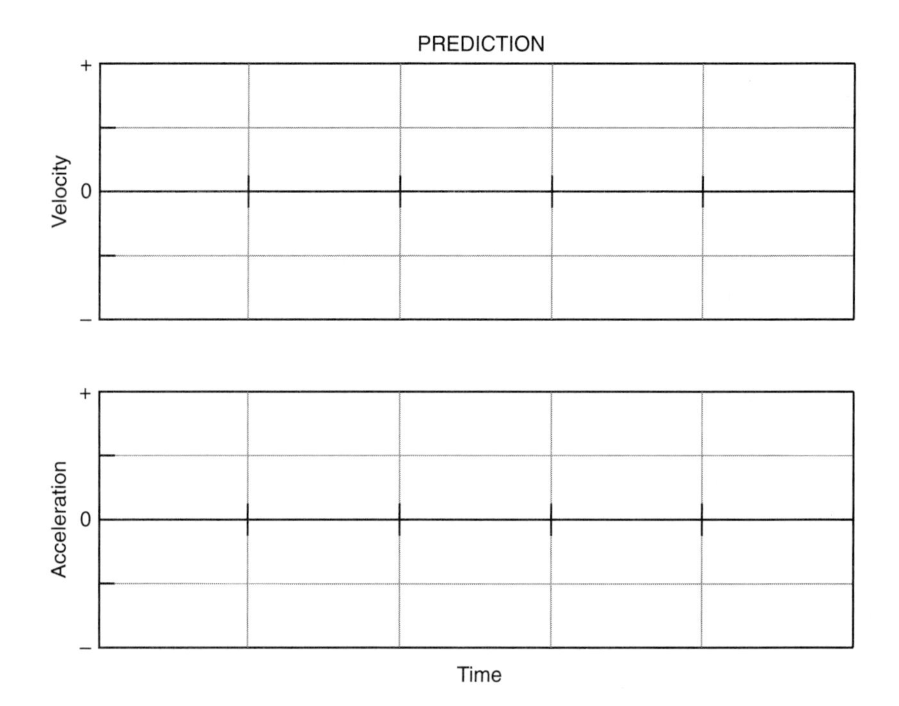

2. Test your predictions. Open the experiment file called **Slowing Down (L02A3-1)** to display the velocity–time and acceleration–time axes that follow.

3. **Begin graphing** with the back of the cart near the 0.5-m mark. Turn the fan unit on, and when you begin to hear the clicks from the motion detector, give the cart a gentle push away from the detector so that it comes to a stop near the end of the ramp. *(Be sure that your hand is not between the cart and the detector.) Stop the cart with your hand—do not let it return toward the motion detector—and turn the fan unit off immediately to save the batteries.*

   You may have to try a few times to get a good run. Don't forget to **change the axes** if this will make your graphs easier to read.

   Move your data so that the graphs are **persistently displayed on the screen.**

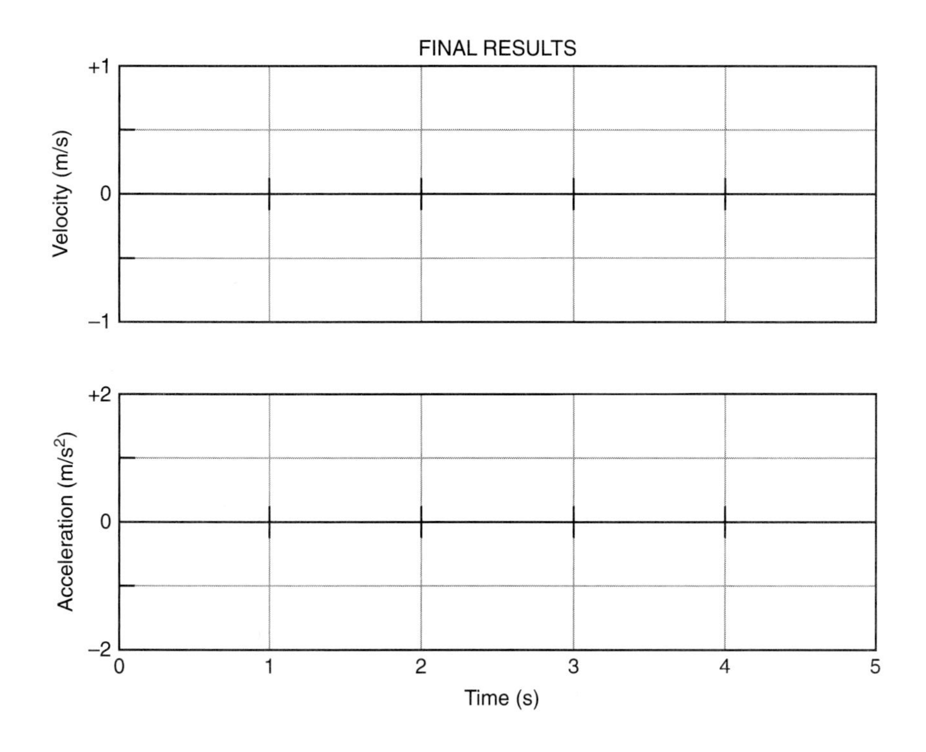

4. Neatly sketch your results on the previous axes, or **print** the graphs and affix them over the axes.

Label your graphs with

- A at the spot where you started pushing.
- B at the spot where you stopped pushing.
- C the region where only the force of the fan is acting on the cart
- D at the spot where the cart came to rest (and you stopped it with your hand).

Also sketch on the same axes the velocity and acceleration graphs for *Speeding Up 2* from Activity 1-2.

Question 3-1: Did the shapes of your velocity and acceleration graphs agree with your predictions? How can you tell the sign of the acceleration from a velocity–time graph?

Question 3-2: How can you tell the sign of the acceleration from an acceleration–time graph?

Question 3-3: Is the sign of the acceleration (which indicates its direction) what you predicted? How does *slowing down* while moving *away* from the detector result in this sign of acceleration? (**Hint:** Remember that acceleration is the *rate of change* of velocity with respect to time. Look at how the velocity is changing.)

Question 3-4: The diagram below shows the positions of the cart at equal time intervals. (This is like overlaying snapshots of the cart at equal time intervals.) At each indicated time, sketch a vector above the cart that might represent the velocity of the cart at that time while it is moving away from the motion detector and slowing down. Assume that the cart is moving at t_1 and t_4.

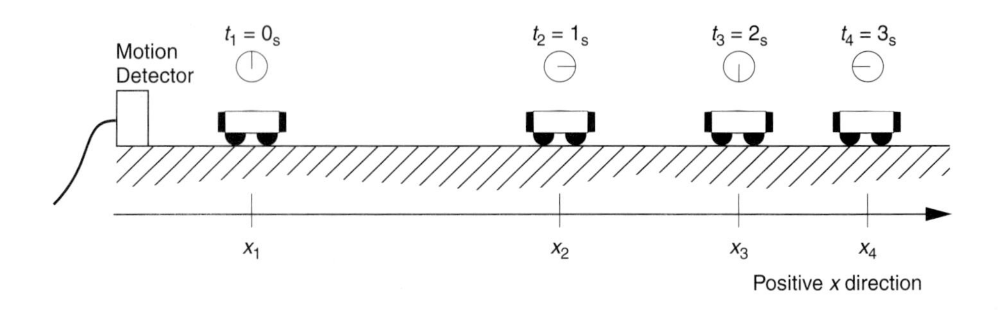

Question 3-5: Show below how you would find the vector representing the change in velocity between the times 2 and 3 s in the diagram above. (Remember that the change in velocity is the final velocity minus the initial velocity.)

Based on the direction of this vector and the direction of the positive x axis, what is the sign (the direction) of the acceleration? Does this agree with your answer to Question 3-3?

Question 3-6: Based on your observations in this lab, state a general rule to predict the sign (the direction) of the acceleration if you know the sign of the velocity (i.e., the direction of motion) and whether the object is speeding up or slowing down.

Activity 3-2 Speeding Up Toward the Motion Detector

Prediction 3-2: Suppose now that you start with the cart at the far end of the ramp, and let the fan push it *toward* the motion detector. As the cart moves toward the detector and speeds up, predict the direction of the acceleration. Will the sign (direction) of the acceleration be positive or negative? (Use your general rule from Question 3-6.)

Sketch your predictions for the velocity–time and acceleration–time graphs on the axes that follow.

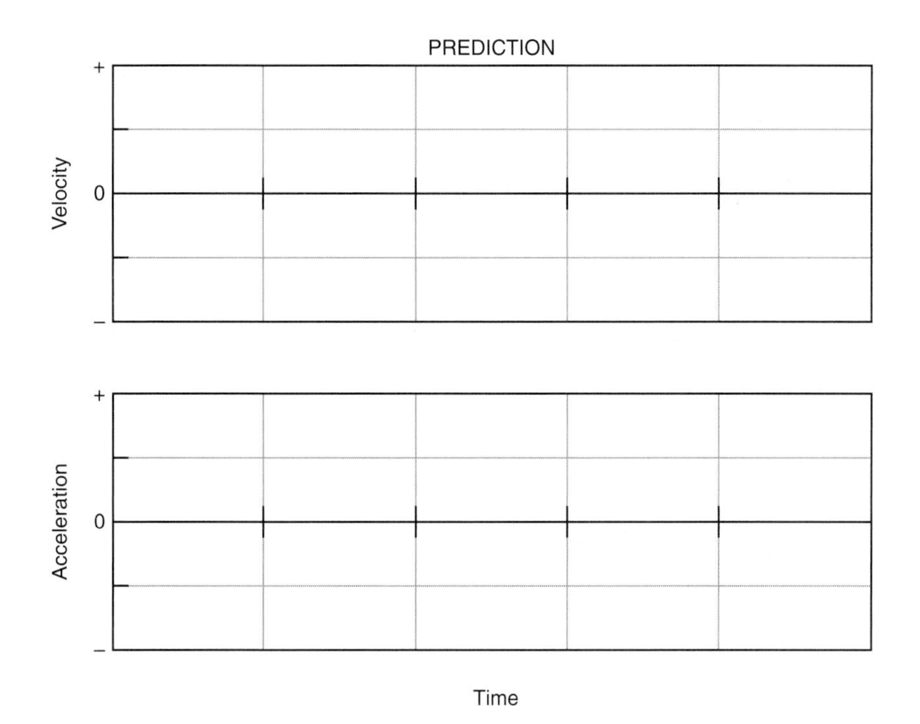

1. Test your predictions. First **clear** any previous graphs. **Graph velocity first.** Graph the cart moving *toward* the detector and *speeding up.* Turn the fan unit on, and when you hear the clicks from the motion detector, release the cart from rest from the far end of the ramp. *(Be sure that your hand is not between the cart and the detector.) Stop the cart when it reaches the 0.5-m line, and turn the fan unit off immediately.*

2. Sketch these graphs or **print** and affix on the axes below. Label these graphs as *"Speeding Up Moving Toward."*

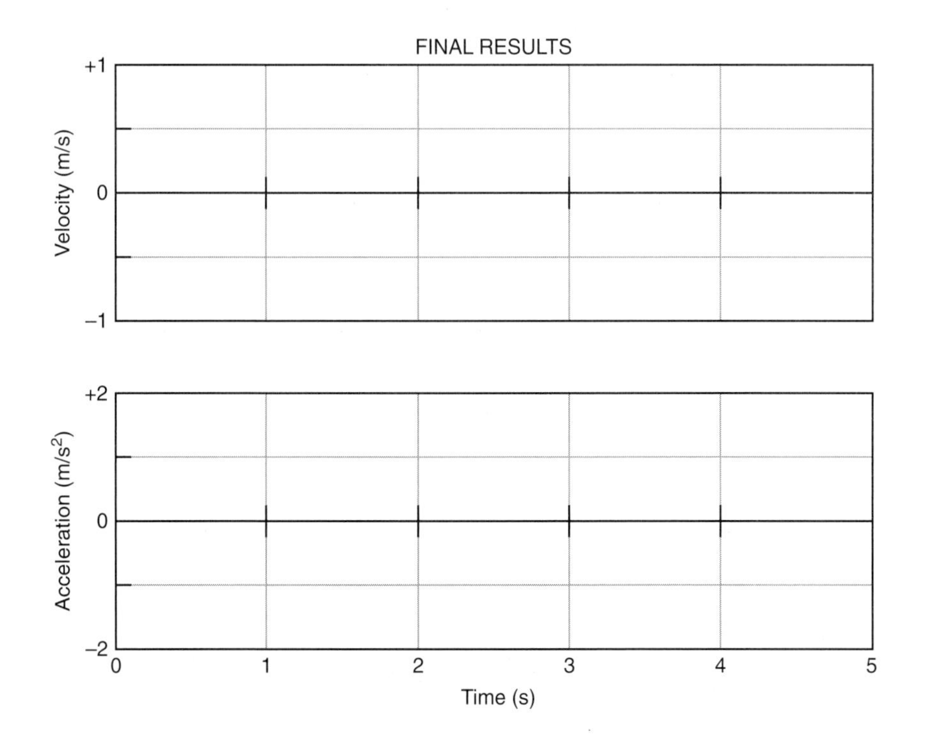

Question 3-7: How does your velocity graph show that the cart was moving *toward* the detector?

Question 3-8: During the time that the cart was speeding up, is the acceleration positive or negative? Does this agree with your prediction? Explain how *speeding up* while moving *toward* the detector results in this sign of acceleration. (**Hint:** Look at how the velocity is changing.)

Question 3-9: When an object is speeding up, what must be the direction of the acceleration relative to the direction of object's velocity? Are they in the same or different directions? Explain.

Question 3-10: The diagram shows the positions of the cart at equal time intervals. At each indicated time, sketch a vector above the cart that might represent the velocity of the cart at that time while it is moving toward the motion detector and speeding up. Assume that the cart is already moving at t_1.

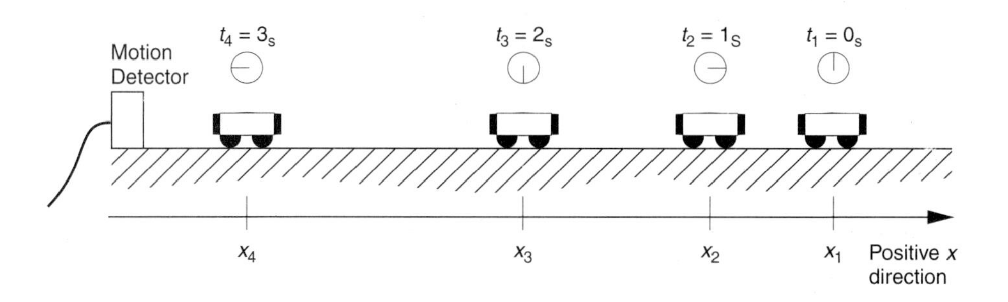

Question 3-11: Show below how you would find the vector representing the change in velocity between the times 2 and 3 s in the diagram above. Based on the direction of this vector and the direction of the positive x axis, what is the sign of the acceleration? Does this agree with your answer to Question 3-8?

Question 3-12: Was your general rule in Question 3-6 correct? If not, modify it and restate it here.

Question 3-13: There is one more possible combination of velocity and acceleration directions for the cart: moving *toward* the detector and *slowing down.* Use your general rule to predict the direction and sign of the acceleration in this case. Explain why the acceleration should have this direction and this sign in terms of the sign of the velocity and how the velocity is changing.

Question 3-14: The diagram shows the positions of the cart at equal time intervals for the motion described in Question 3-13. At each indicated time, sketch a vector above the cart that might represent the velocity of the cart at that time while it is moving toward the motion detector and slowing down. Assume that the cart is moving at t_1 and t_4.

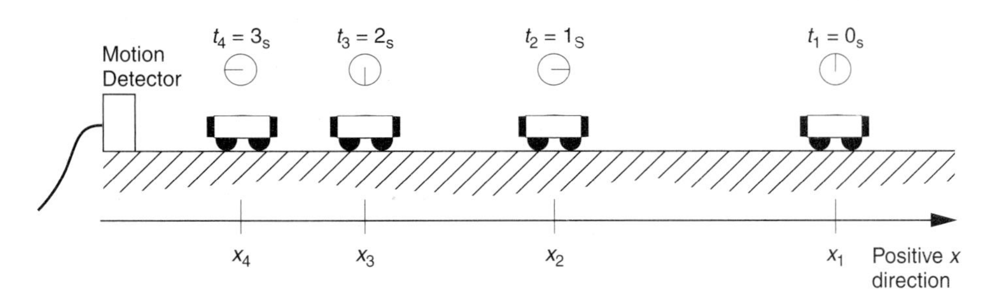

Question 3-15: Show how you would find the vector representing the change in velocity between the times 2 and 3 s in the diagram above. Based on the direction of this vector and the direction of the positive x axis, what is the sign of the acceleration? Does this agree with your answer to Question 3-13?

If you have more time, do the following Extension.

Extension 3-3: Graphing Slowing Down Toward the Motion Detector

Use the motion detector setup to graph the motion of the cart moving *toward* the motion detector and *slowing down,* as described in Question 3-13. **Print** and affix the graph.

Question E3-16: Compare the graphs to your answers to Questions 3-13 to 3-15.

Activity 3-4: Reversing Direction

In this activity you will look at what happens when the cart slows down, reverses its direction and then speeds up in the opposite direction. How does the velocity change with time? What is the cart's acceleration?

The setup should be as shown below—the same as before. The fan unit should have the maximum number of batteries, and should be taped securely to the cart.

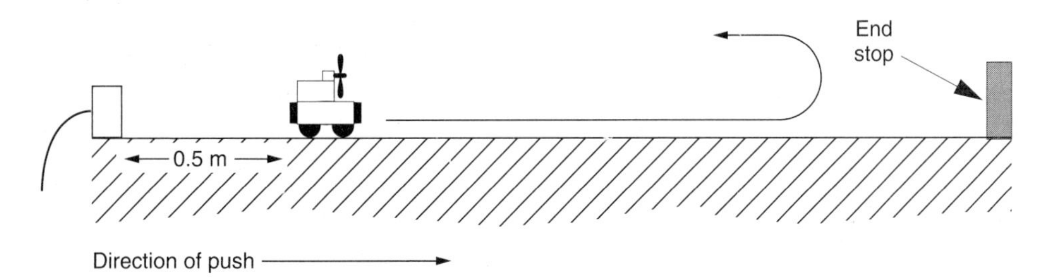

Prediction 3-3: You start the fan and give the cart a push *away* from the motion detector. It moves away, slows down, reverses direction, and then moves back to-

ward the detector. Try it without using the motion detector! *Be sure to stop the cart before it hits the motion detector, and turn the fan off immediately.*

For each part of the motion—*away from the detector, at the turning point,* and *toward the detector*—indicate in the table below whether the velocity is positive, zero, or negative. Also indicate whether the acceleration is positive, zero, or negative.

PREDICTION TABLE

	Moving away	At the turning point	Moving toward
Velocity			
Acceleration			

On the axes that follow sketch your predictions of the velocity–time and acceleration–time graphs of this entire motion.

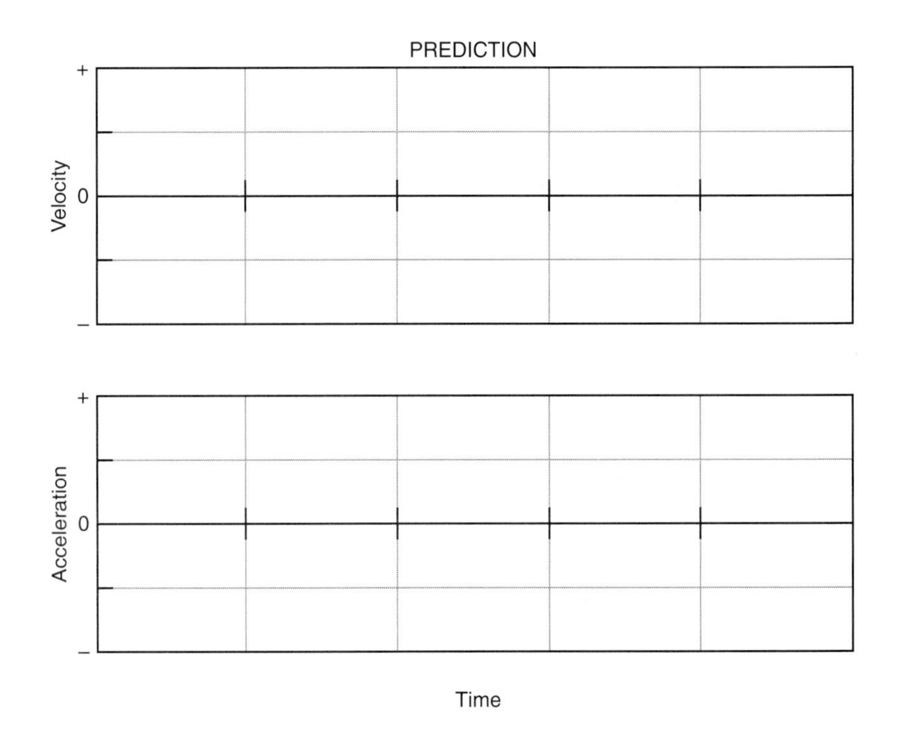

1. Test your predictions. Set up to graph velocity and acceleration on the following graph axes. (Open the experiment file called **Slowing Down (L02A3-1)** if it is not already opened.)

2. **Begin graphing** with the back of the cart near the 0.5-m mark. Turn on the fan unit, and when you begin to hear the clicks from the motion detector, give the cart a gentle push away from the detector so that it travels at least 1 m, slows down, and then reverses its direction and moves toward the detector. *(Push and stop the cart with your hand on its side. Be sure that your hand is not between the cart and the detector.)*

   Be sure to stop the cart at least 0.5-m from the motion detector and turn off the fan unit immediately.

   You may have to try a few times to get a good round trip. Don't forget to change the scales if this will make your graphs clearer.

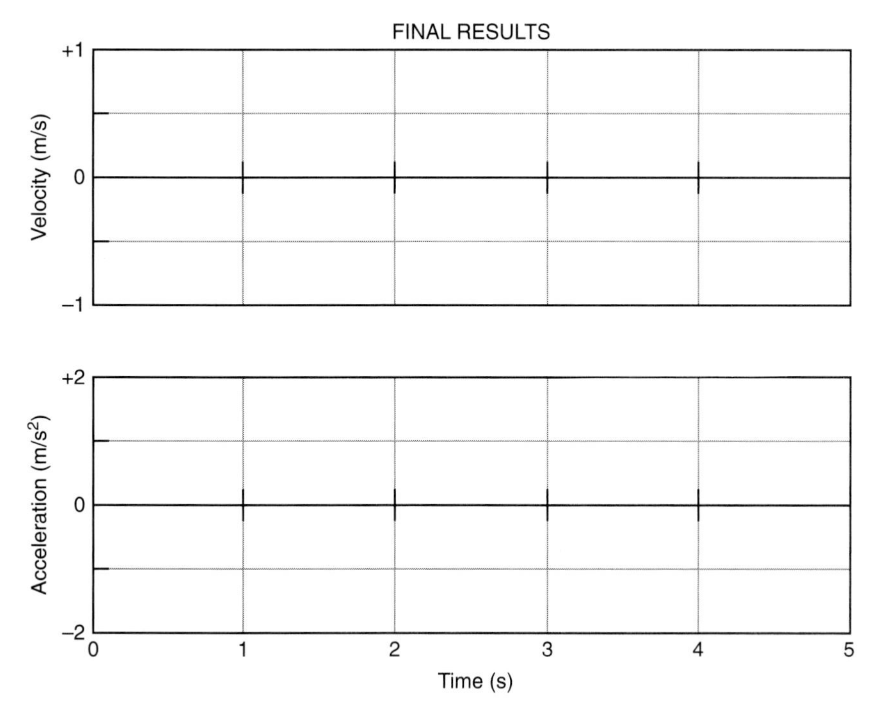

3. When you get a good round trip, sketch both graphs on the axes above or **print** and affix over the axes.

Question 3-17: Label *both* graphs with

- A where the cart started being pushed.
- B where the push ended (where your hand left the cart).
- C where the cart reached its turning point (and was about to reverse direction).
- D where you stopped the cart with your hand.

Explain how you know where each of these points is.

Question 3-18: Did the cart "stop" at its turning point? (**Hint:** Look at the velocity graph. What was the velocity of the cart at its turning point?) Does this agree with your prediction? How much time did it spend at the turning point velocity before it started back toward the detector? Explain.

Question 3-19: According to your acceleration graph, what is the acceleration at the instant the cart reaches its turning point? Is it positive, negative, or zero? Is it significantly different from the acceleration during the rest of the motion? Does this agree with your prediction?

Question 3-20: Explain the observed sign of the acceleration at the turning point. (**Hint:** Remember that acceleration is the *rate of change* of velocity. When the cart is at its turning point, what will its velocity be in the next instant? Will it be positive or negative?)

Question 3-21: On the way back toward the detector, is there any difference between these velocity and acceleration graphs and the ones that were the result of the cart starting from rest (Activity 3-2)? Explain.

If you have more time, do the following Extension.

Extension 3-5: Sign of Push and Stop

Find on your acceleration graphs for Activity 3-4 the time intervals when you pushed the cart to start it moving and when you stopped it.

Question E3-22: What is the sign of the acceleration for each of these intervals? Explain why the acceleration has this sign in each case.

Challenge: You throw a ball up into the air. It moves upward, reaches its highest point, and then moves back down toward your hand. Assuming that upward is the positive direction, indicate in the table that follows whether the velocity is positive, zero, or negative during each of the three parts of the motion. Also indicate if the acceleration is positive, zero, or negative. (**Hint:** Remember that to find the acceleration, you must look at the *change* in velocity.)

	Moving up *after release*	At highest point	Moving down
Velocity			
Acceleration			

Question 3-23: In what ways is the motion of the ball similar to the motion of the cart that you just observed? What causes the ball to accelerate?

Name______________________ Date__________ Partners______________________

Homework for Lab 2: Changing Motion

1. An object moving along a line (the + position axis) has the acceleration–time graph on the right.

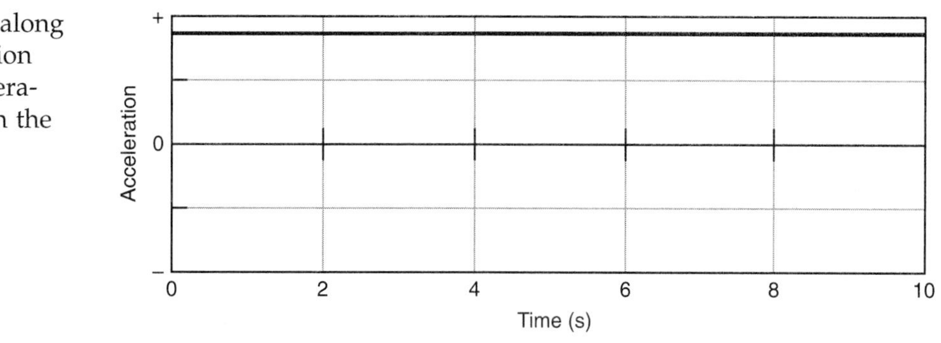

a. Describe how the object might move to create this graph if it is moving away from the origin.

b. Describe how the object might move to create this graph if it is moving toward the origin.

c. Sketch with a solid line on the axes on the right a velocity–time graph that goes with the motion described in (a).

d. Sketch with a dashed line on the axes on the right a velocity–time graph that goes with the motion described in (b).

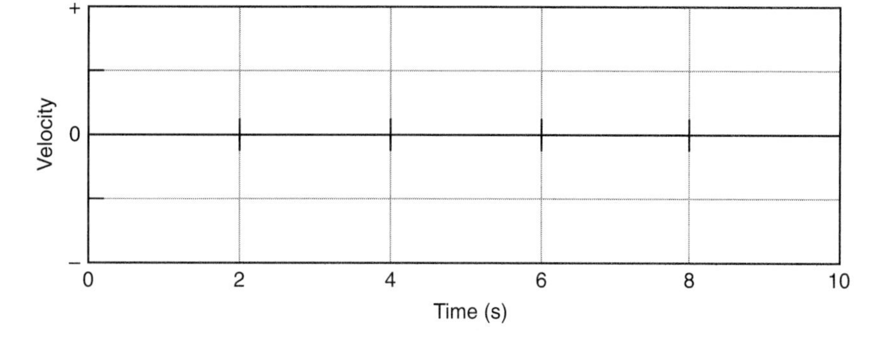

2. How would an object move to create each of the three labeled parts of the acceleration–time graph shown below?

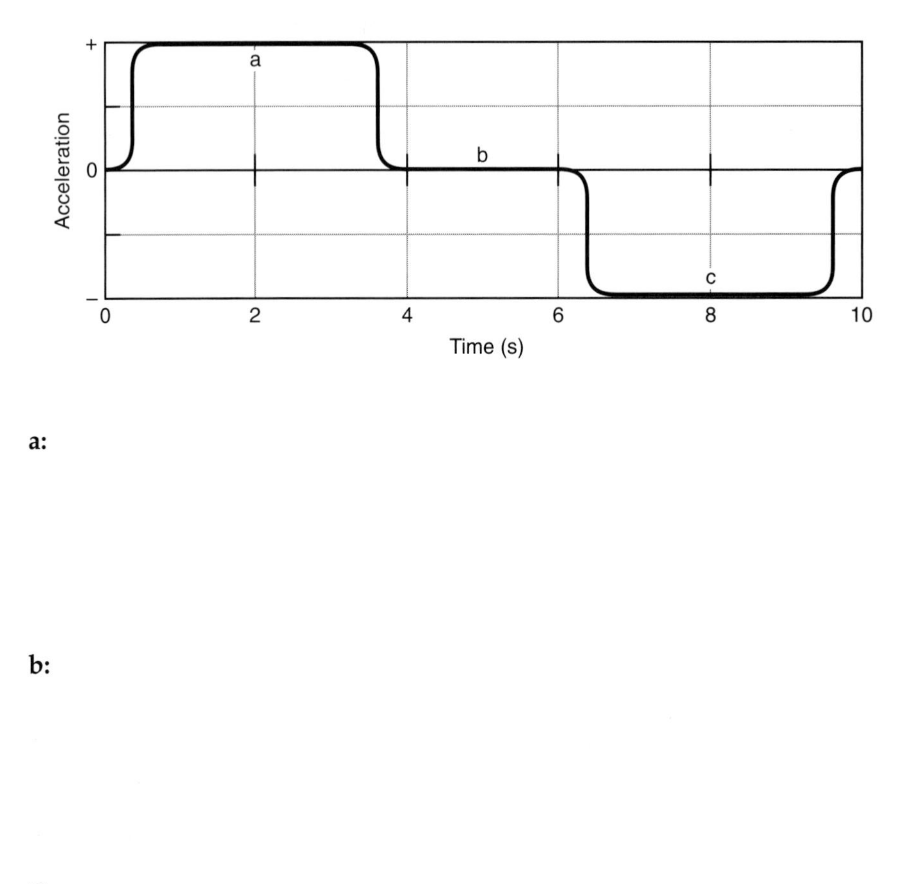

a:

b:

c:

3. Sketch below a velocity–time graph that might go with the acceleration–time graph in Question 2.

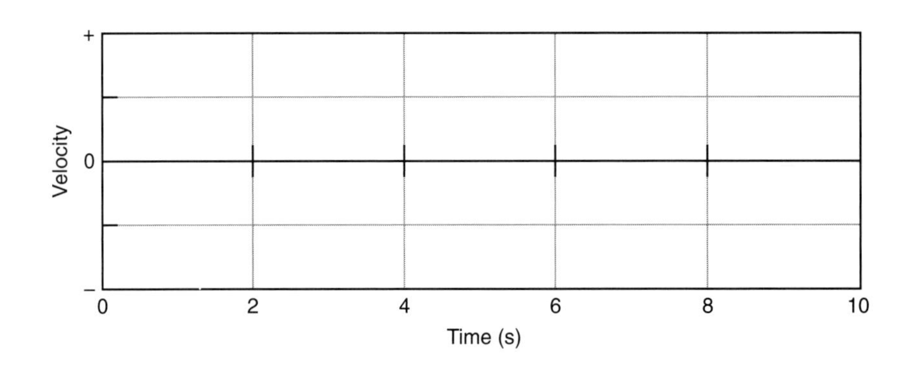

4. For each of the velocity–time graphs that follow, sketch the shape of the acceleration–time graph that goes with it.

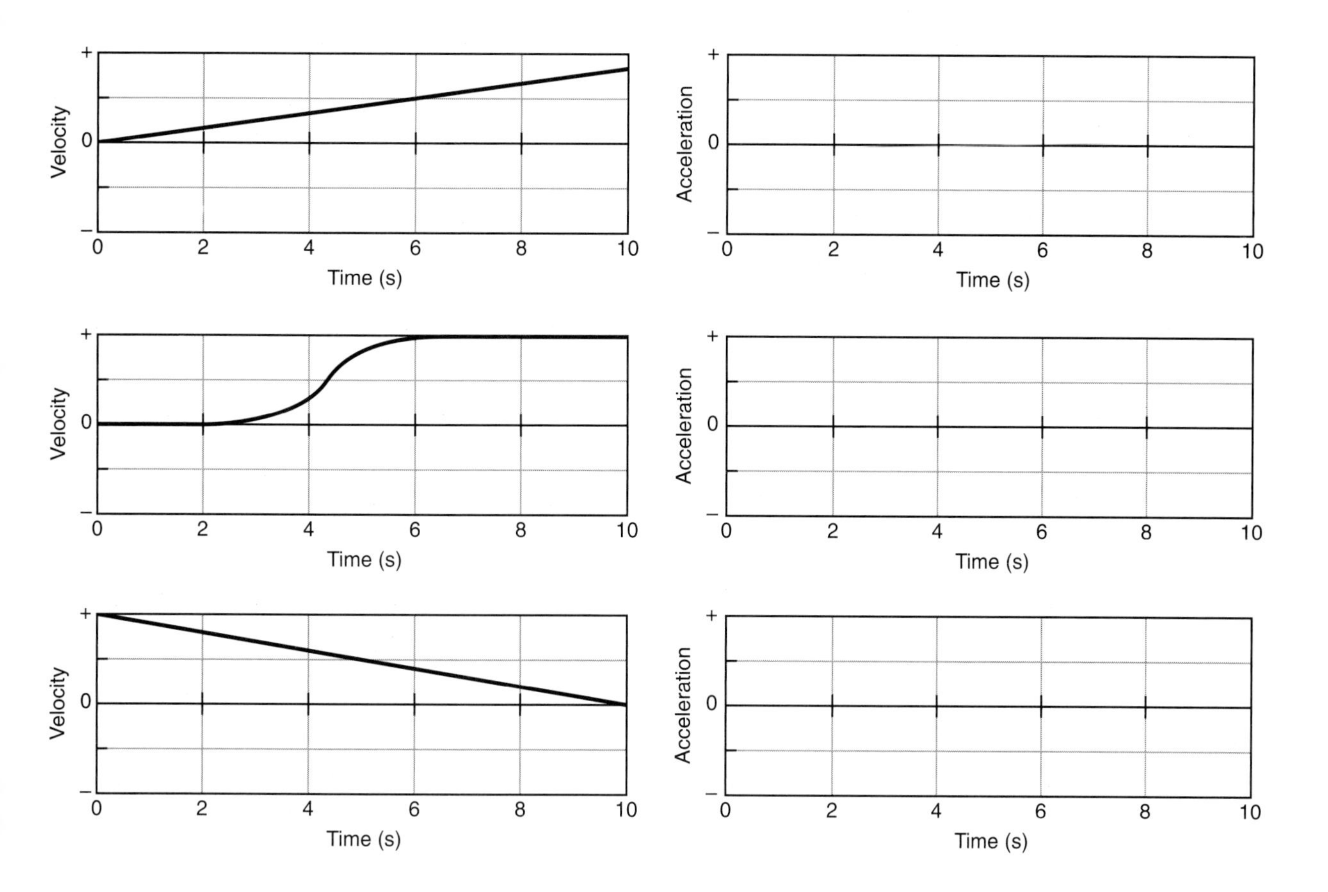

5. A car can move along a line (the + position axis). Sketch velocity–time and acceleration–time graphs that correspond to each of the following descriptions of the car's motion.

   a. The car starts from rest, and moves away from the origin, increasing its speed at a steady rate.

b. The car is moving away from the origin at a constant velocity.

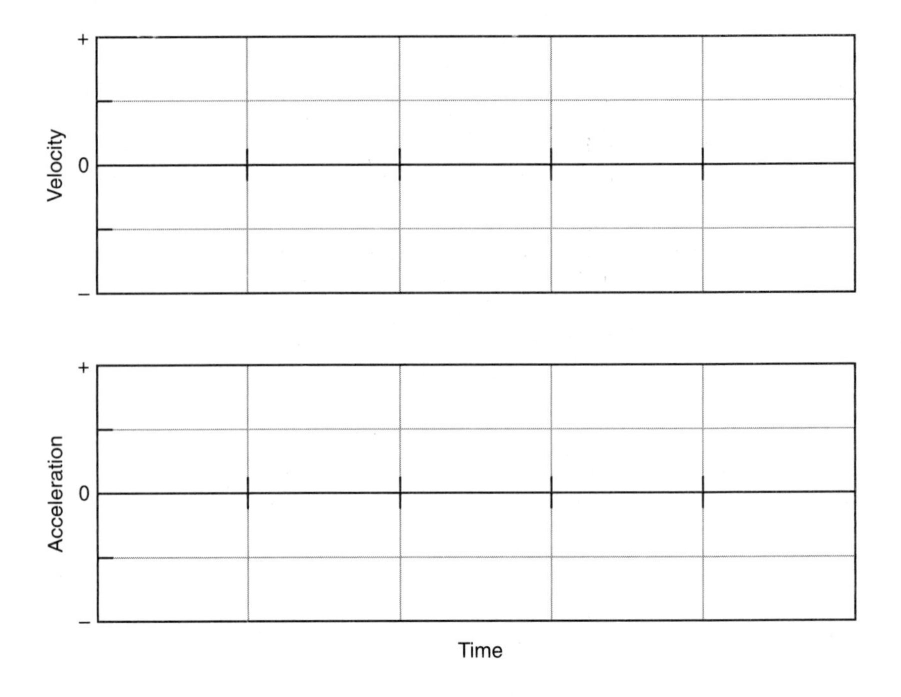

c. The car starts from rest, and moves away from the origin, increasing its speed at a steady rate twice as large as in (a) above.

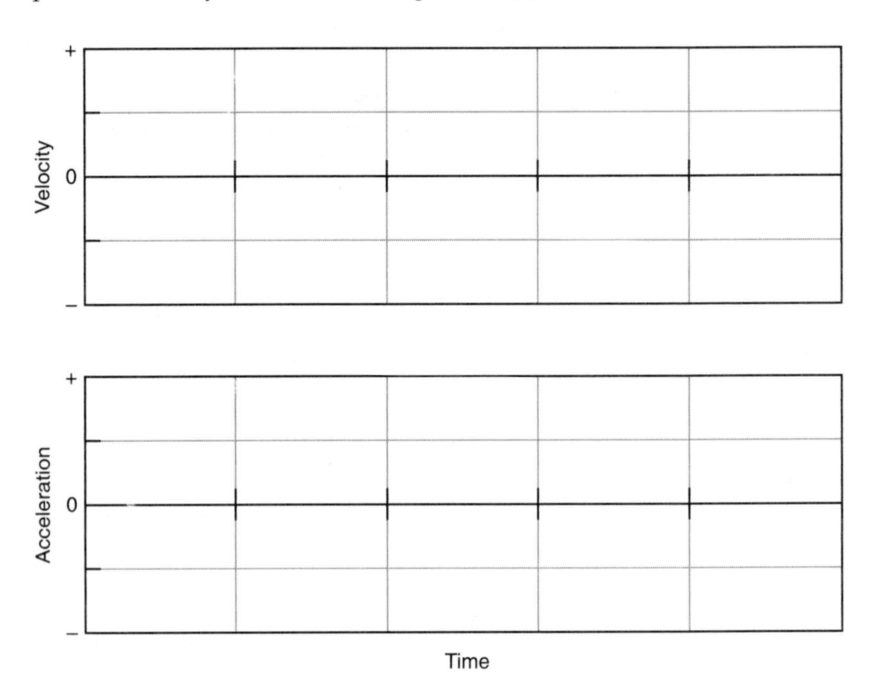

d. The car starts from rest, and moves toward the origin, increasing its speed at a steady rate.

+
Velocity 0
–

+
Acceleration 0
–

Time

e. The car is moving toward the origin at a constant velocity.

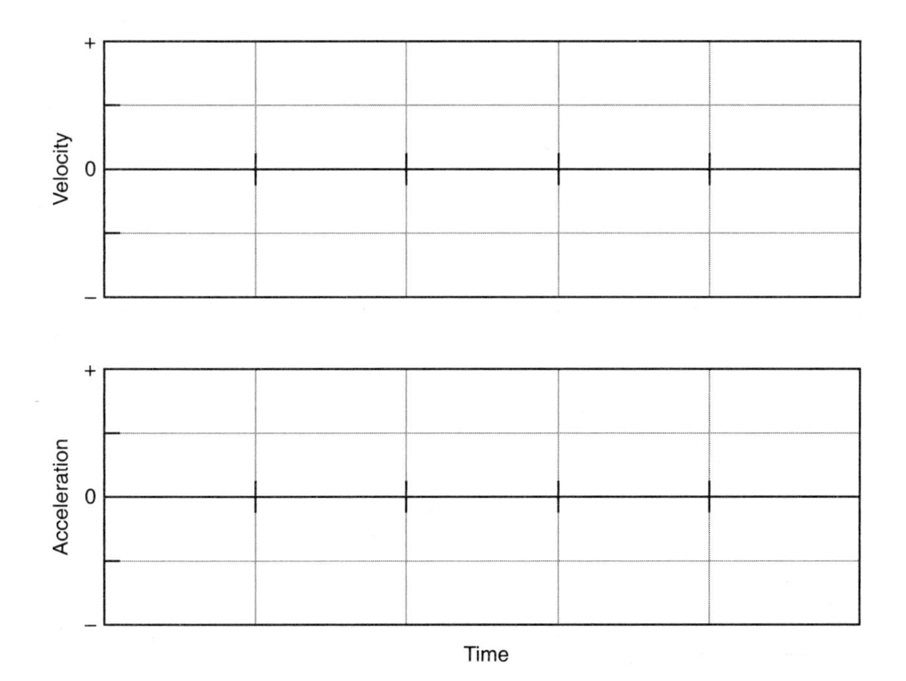

6. The following is a velocity–time graph for a car.

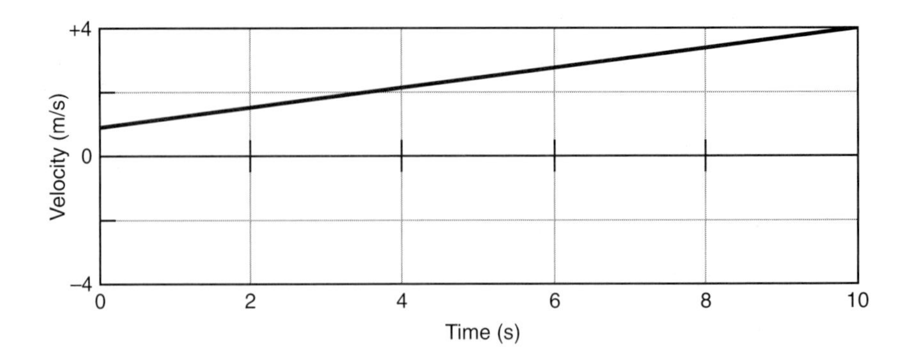

What is the average acceleration of the car? Show your work below.

7. A car moves along a line (the + position axis). Fill in the table below with the sign (+ or −) of the velocity and acceleration of the car for each of the motions described.

	Position	Velocity	Acceleration when speeding up	Acceleration when slowing down
Car moves away from the origin	+			
Car moves toward the origin	+			

8. A ball is tossed in the air. It moves upward, reaches its highest point, and falls back downward. Sketch a velocity–time and an acceleration–time graph for the ball from the moment it leaves the thrower's hand until the moment just before it reaches her hand again. Consider the positive direction to be *upward*.

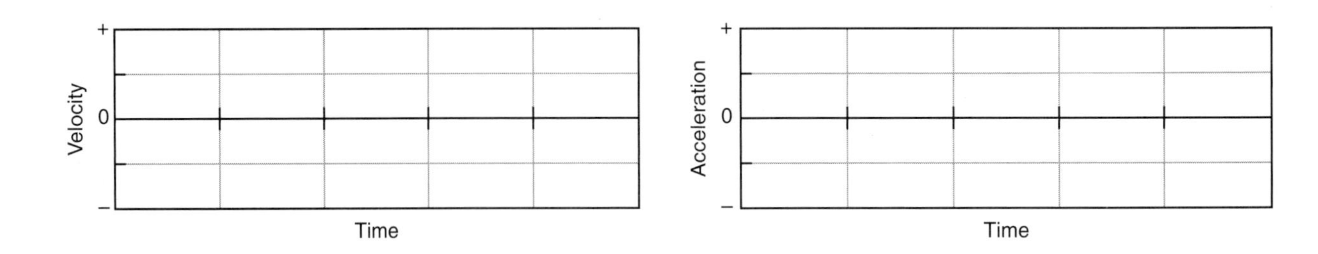

9. For each of the position–time graphs shown, sketch below it the corresponding velocity–time and acceleration–time graphs.

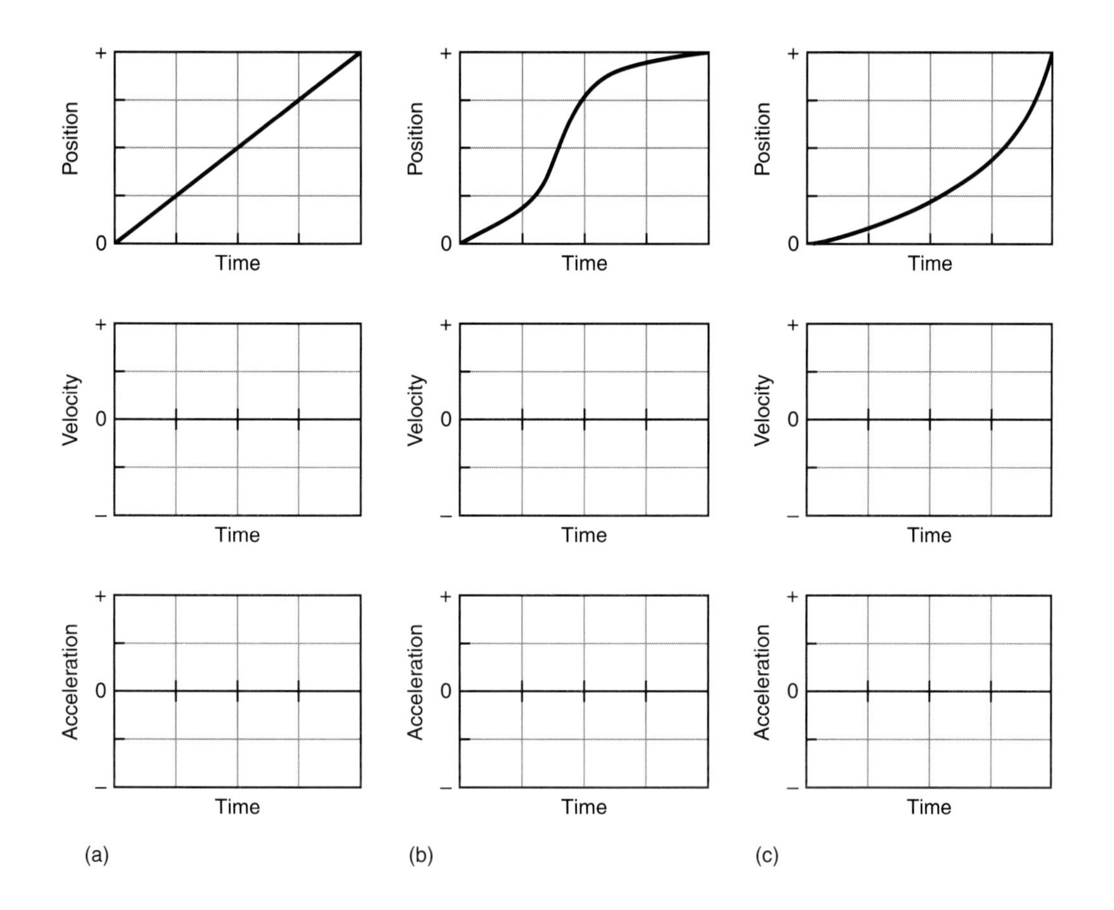

10. Each of the pictures below represents a car moving down a road. The motion of the car is described. In each case, draw velocity and acceleration vectors above the car that might represent the described motion. Label the vectors. Also specify the sign of the velocity and the sign of the acceleration. (The positive position direction is toward the right.)

a. The driver has stepped on the accelerator, and the car is just starting to move forward.

Sign of velocity:

Sign of acceleration:

b.

The car is moving forward. The brakes have been applied. The car is slowing down, but has not yet come to rest.

Sign of velocity:

Sign of acceleration:

c. The car is moving backward. The brakes have been applied. The car is slowing down, but has not yet come to rest.

Sign of velocity:

Sign of acceleration:

11. **a.** Describe how you would move to produce the velocity–time graph below.

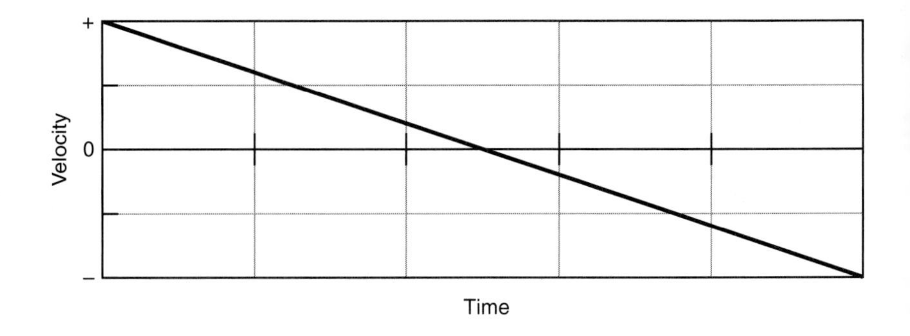

b. Sketch a position–time graph for this motion on the axes below.

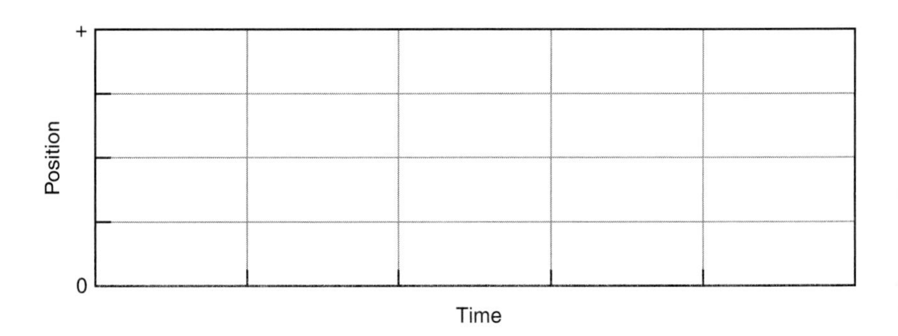

c. Sketch an acceleration–time graph for this motion on the axes below.

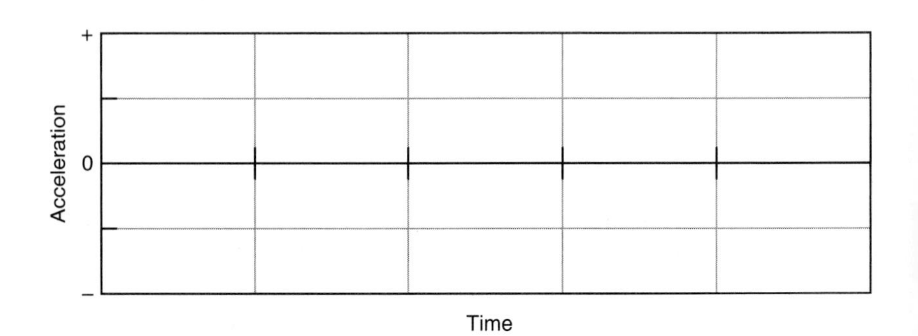

The graphs on this page represent the motion of objects along a line that is the positive position axis. Notice that the motion of objects is represented by position, velocity, or acceleration graphs.

Answer the following questions. You may use a graph more than once or not at all, and there may be more correct choices than blanks. If none of the graphs is correct, answer J.

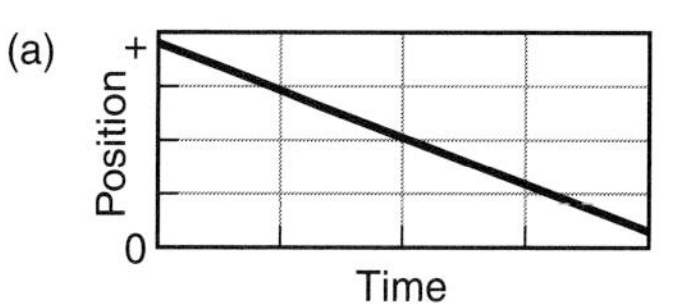

______**12.** Pick one graph that gives enough information to indicate that the velocity is always negative.

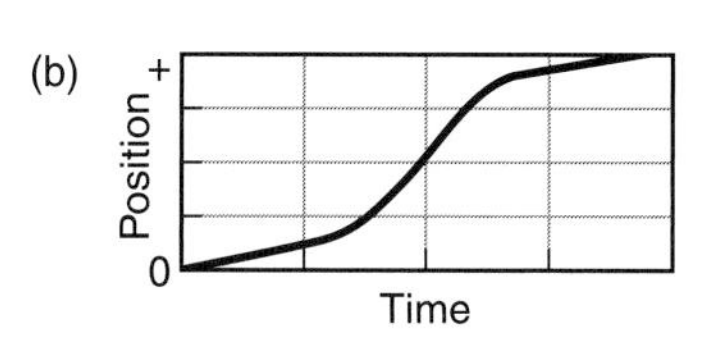

Pick three graphs that represent the motion of an object whose velocity is constant (not changing).

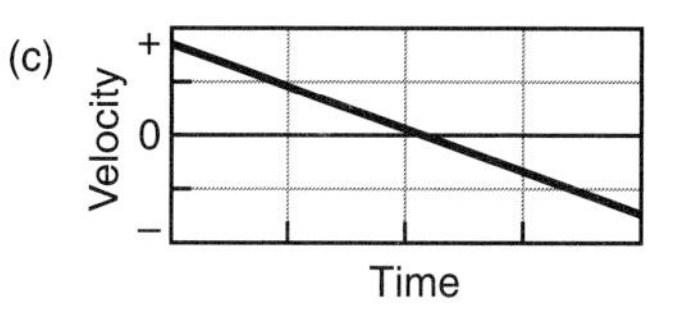

______**13.** ______**14.** ______**15.**

______**16.** Pick one graph that *definitely indicates* an object has reversed direction.

______**17.** Pick one graph that might possibly be that of an object standing still.

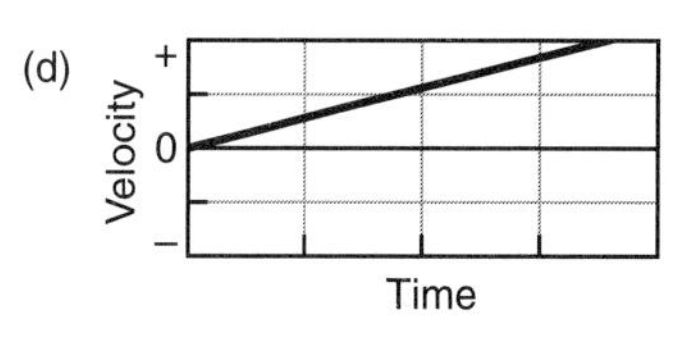

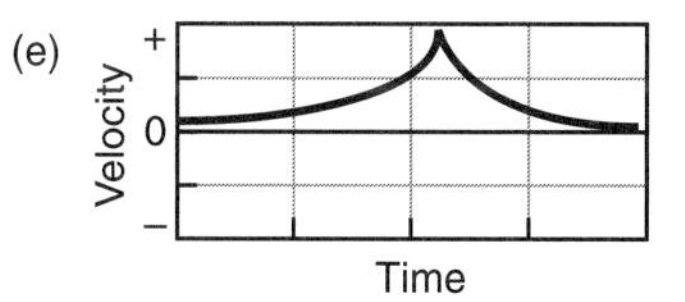

Pick 3 graphs that represent the motion of objects whose acceleration is changing.

______**18.** ______**19.** ______**20.**

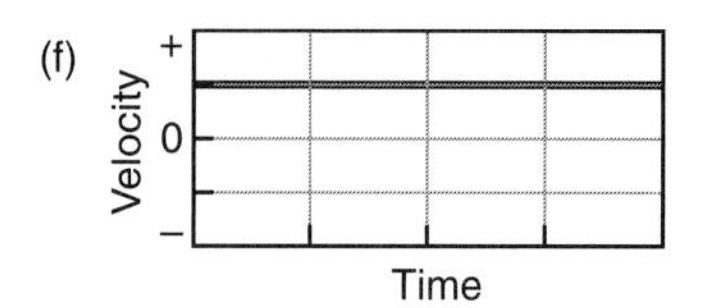

Pick a velocity graph and an acceleration graph that could describe the motion of the same object during the time shown.

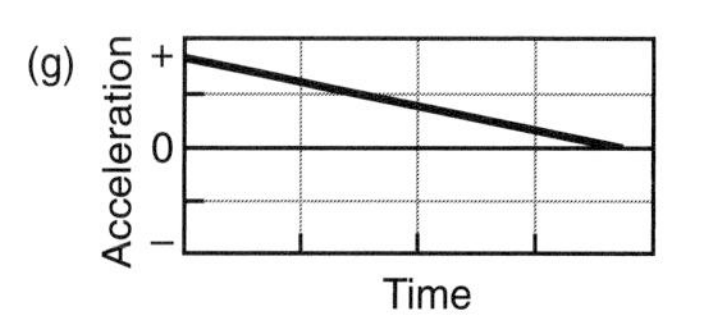

______**21.** Velocity graph.

______**22.** Acceleration graph.

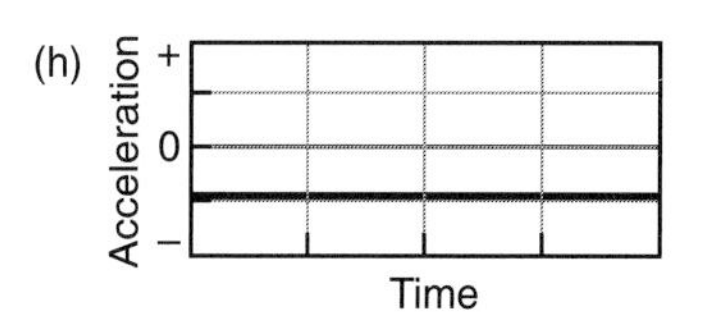

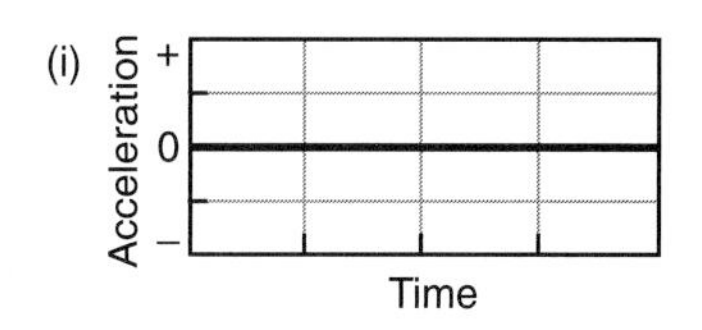

Name____________________________________ Date____________________

Pre-Lab Preparation Sheet for Lab 3: Force and Motion

(Due at the beginning of Lab 3)

Directions:
Read over Lab 3 and then answer the following questions about the procedures.

1. What is the purpose of the rubber bands in Activity 1-1?

2. What is the difference between a *linear* and a *proportional* relationship?

3. Why is it necessary to calibrate a force probe?

4. Sketch your Prediction 2-2 below.

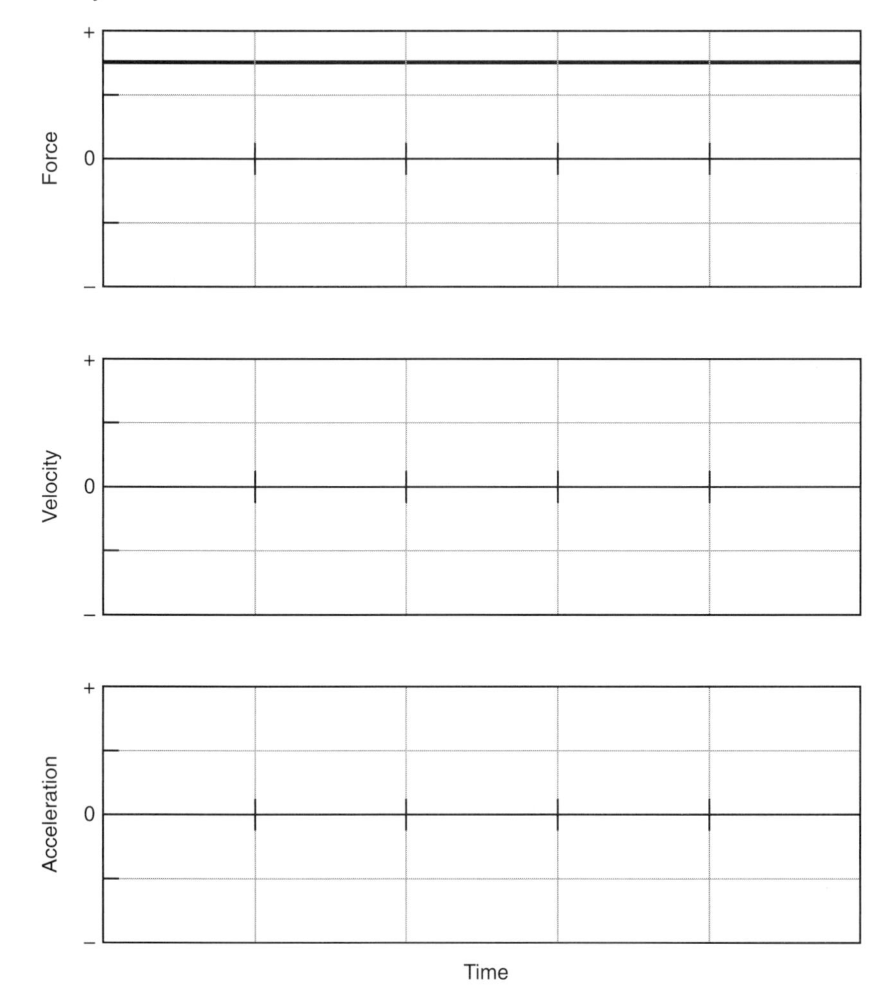

Name______________________ Date____________ Partners______________________

LAB 3: FORCE AND MOTION

A vulgar Mechanik can practice what he has been taught or seen done, but if he is in an error he knows not how to find it out and correct it, and if you put him out of his road, he is at a stand; whereas he that is able to reason nimbly and judiciously about figure, force and motion, is never at rest til he gets over every rub.

—Isaac Newton

OBJECTIVES

- To develop a method for measuring forces reliably.
- To learn how to use a force probe to measure force.
- To explore how the motion of an object is related to the forces applied to it.
- To find a mathematical relationship between the force applied to an object and its acceleration.

OVERVIEW

In the previous labs, you have used a motion detector to display position–time, velocity–time, and acceleration–time graphs of different motions of various objects. You were not concerned about how you got the objects to move, i.e., what forces (pushes or pulls) acted on the objects. From your own experiences, you know that force and motion are related in some way. To start your bicycle moving, you must apply a force to the pedal. To start up your car, you must step on the accelerator to get the engine to apply a force to the road through the tires.

But exactly how is force related to the quantities you used in the previous lab to describe motion—position, velocity, and acceleration? In this lab you will pay attention to forces and how they affect motion. You will first develop an idea of a force as a push or a pull. You will learn how to measure forces. By applying forces to a cart and observing the nature of its resulting motion graphically with a motion detector, you will come to understand the effects of forces on motion.

INVESTIGATION 1: MEASURING FORCES

In this investigation you will explore the concept of a constant force and the combination of forces in one dimension. You can use these concepts to learn how to measure forces with a force probe. You will need the following materials:

- computer-based laboratory system
- *RealTime Physics Mechanics* experiment configuration files
- force probe
- five identical rubber bands
- meter stick
- spring scale with a maximum reading of 5 N

Activity 1-1: How Large Is a Pull?

If you pull on a rubber band attached at one end, you know it will stretch. The more you pull, the more it stretches. Try it.

1. Attach one end of the rubber band to something on the table that can't move. Also attach the meter stick to the table. Now stretch the rubber band so it is several centimeters longer than its relaxed length. Does it always seem to exert the *same* pull on you each time it is stretched to the *same* length? (Most people agree that this is obvious.)
2. Write down the length you have chosen in the space below. This will be your *standard length* for future measurements.

   Standard length of rubber band = ______cm

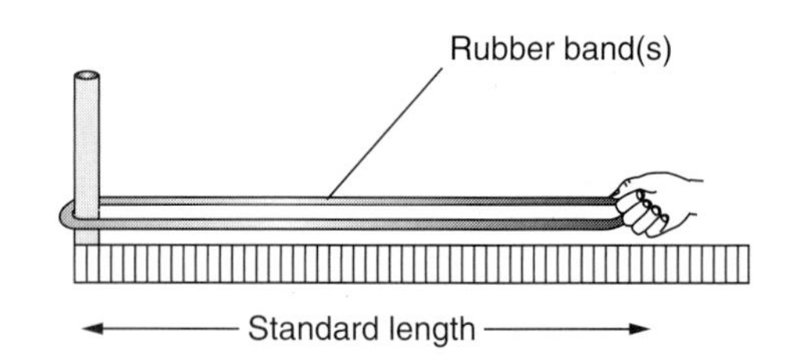

3. Attach one end of each of two identical rubber bands to something that can't move and stretch them together side-by-side to the standard length.

Question 1-1: How does the combined force of two rubber bands compare to what you felt when only one rubber band was used?

4. Repeat this comparison of how strong the forces feel with three, four, and five rubber bands stretched together to the same standard length.

Question 1-2: Suppose you stretched a rubber band to your standard length by pulling on it. Now you want to create a force six times as large. How could you create such a force?

Question 1-3: Suppose you applied a force with a stretched rubber band one day, and several days later you wanted to feel the same force or apply it to something. How could you assure that the forces were the same? Explain.

Question 1-4: Do side-by-side rubber bands provide a convenient way of accurately reproducing forces of many different sizes? Explain.

You have seen that pulling more rubber bands to the same length requires a larger pull. To be more precise about the pulls and pushes you are applying, you need a device to measure forces accurately. The electronic force probe is designed to do this.

Activity 1-2: Measuring Forces With a Force Probe

In this activity you will explore the capability of an electronic force probe as a force measuring device.

1. Plug the force probe into the computer interface. Display force vs. time axes by opening the experiment file called **Measuring Force (L03A1-2a).**

   If the sensitivity of your force probe can be adjusted, choose the setting appropriate to the measurements in this activity (about 0 to 10 N or more).

> **Comment:** Since forces are detected by the computer system as *changes* in an electronic signal, it is important to have the computer "read" the signal when the force probe has no force pushing or pulling on it. This process is called "zeroing." Also, the electronic signal from the force probe can change slightly from time to time as the temperature changes. This is especially true with a Hall effect force probe. Therefore, if it is possible to **zero** your force probe, it is a good idea to do so with nothing attached to the probe before making each measurement.

2. If possible, **zero** the force probe with nothing pulling on it.
3. Attach one end of a rubber band to something that can't move, as before, and the other end to the force probe hook. Pull horizontally on the rubber band with the force probe until the band is stretched to the same *standard length* used in Activity 1-1, and **begin graphing** while holding the rubber band steady for the whole graph.

   Record a typical force probe reading: ______

   (Use the **analysis feature** of the software to get an accurate reading of the force.)

4. Although you could continue to take data with multiple rubber bands in the same way you did with just one, the easiest way to take these data is to open the experiment file called **Rubber Bands (L03A1-2b).** This experiment sets the software in **event mode** and allows you to plot the measured force against the number of rubber bands by manually entering the number when prompted.

   After loading, **zero** the force probe and then **begin graphing.** The force probe reading will be displayed on the graph continuously until you decide to **keep** the value.

   Keep the force values with 0, 1, 2, 3, 4, and 5 rubber bands pulled by the force probe to your standard length. **Stop** graphing when you are done.

5. Record the force probe reading for each number of rubber bands in the table below. (You may use the **analysis feature** of the software to get accurate force readings.)

Number of rubber bands	0	1	2	3	4	5
Force probe reading (arbitrary units)						

Comment: We are interested in the nature of the mathematical relationship between the reading of the force probe and force (in rubber band units). This can be determined from the graph by drawing a smooth curve that fits the plotted data points. Some definitions of possible mathematical relationships are shown below. In these examples, y might be the force probe reading and x the number of rubber bands.

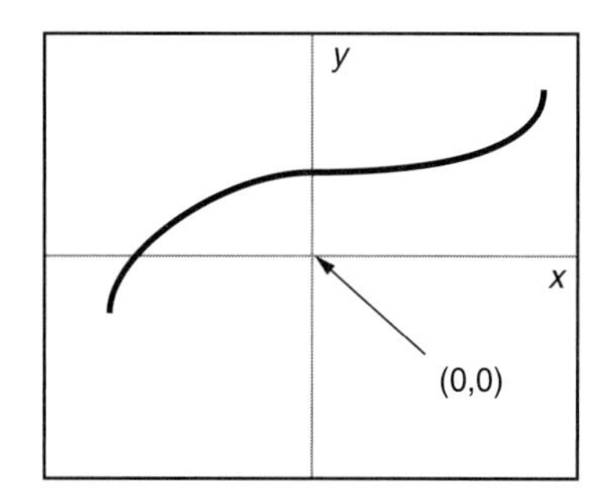

y is a function of x, which increases as x increases.

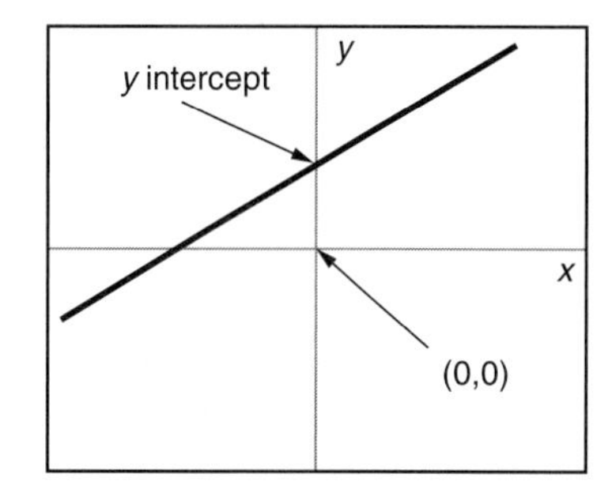

y is a *linear* function of x, which increases as x increases according to the mathematical relationship $y = mx + b$, where b is a constant called the *y intercept.*

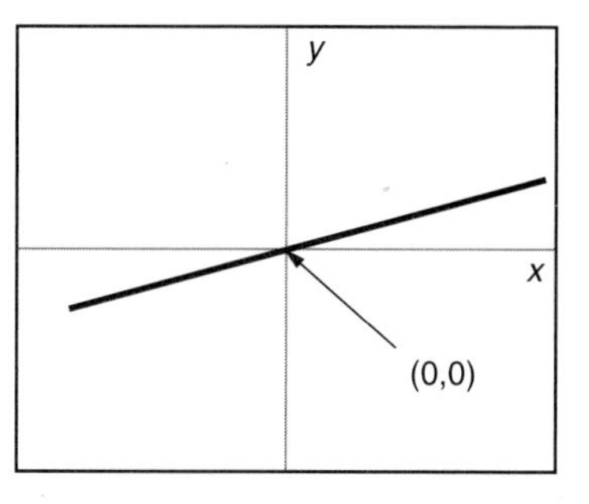

y is *proportional* to x. This is a special case of a linear relationship where $y = mx$, and b, the *y intercept,* is zero.

These graphs show the differences between these three types of mathematical relationship. y can increase as x increases, and the relationship doesn't have to be *linear* or *proportional. Proportionality refers only to the special linear relationship where the y intercept is zero, as shown in the example graph on the right.*

6. The **fit routine** of your software allows you to determine the relationship by trying various curves to see which best fits the plotted data. Try a linear fit: $F = c + bx$ and any other that you think is appropriate.

When you have found a good fit, where the curve goes through or close to all of the data points, **print** the graph along with the fit equation, and affix it in the space below.

Question 1-5: How are force probe readings related to the size of the pull exerted by the force probe hook on the rubber bands? Describe the mathematical relationship in words.

Question 1-6: Based on your graph, what force probe reading corresponds to the pull of one rubber band when stretched to your standard length? How did you determine this?

Comment: You can use your measurements to define a quantitative force scale. You might call it the "rubber band scale," or give it yours or your partner's name. Whenever the force probe has the reading corresponding to the pull of one rubber band stretched the standard length, the force is equivalent to one "rubber band," or one "Mary" or one "Sam." Any larger force can be measured as some number of these units.

Your graph relates two different ways of measuring force, one with standard stretches of different numbers of rubber bands (rubber band units) and the other with a force probe. Such a graph is called a *calibration curve* and is used to compare measurements of quantities made with two different measuring instruments. You could use it to convert forces measured in force probe units to rubber band units, and vice versa.

Physicists have defined a standard unit of force called the *newton,* abbreviated N. For the rest of your work on forces and the motions they cause, it will be more convenient to have the force probe read directly in newtons. Then the forces you measure can be compared to forces anyone else measures. Most spring scales have already been calibrated in newtons. All you need to do is to calibrate the force probe to read forces in newtons by using the spring scale to input standard force measurements.

If you have extra time, do the following Extension to examine the importance of a linear force scale like that of the spring scale.

Extension 1-3: Importance of Linear Force Scales

You have probably observed in Activity 1-2 that a force scale based on the force probe is approximately a proportional one. That is, if you apply twice as large a force (e.g., by pulling two side-by-side rubber bands rather than one to the standard length) the force probe gives just about twice the reading. Not all devices give a proportional force scale.

Use the force probe to examine the force scales produced by (1) a single rubber band and (2) a spring, each stretched to various lengths. That is, set up experiments to graph the force applied as a function of length for each of these.

Question E1-7: Based on your graphs, is either of these force scales proportional? What are the advantages of using a proportional force scale?

Activity 1-4: Calibrating the Force Probe

Comment: Most strain gauge type force probes have a good calibration that can just be loaded. If you are using a magnetic (Hall effect) type force probe, the sensitivity changes as the spacing between the magnet and the sensor is changed. The smaller the spacing, the more sensitive it is. Be sure to **check the sensitivity** with a 2.0-N force applied, and reset the spacing if necessary.

1. Open the experiment file called **Calibrating Force (L03A1-4).**
2. **Calibrate** the force probe, by following the directions in the **calibration routine** exactly, or **load the calibration.**

   When the computer asks you to apply a known force to the force probe, pull on the hook with the spring scale, holding it with a steady 2.0-N reading on the scale.

   When the calibration is complete, you may release the spring scale.
3. Check the calibration. First **zero** the force probe. Then **begin graphing,** and pull on the force probe with the spring scale with several different forces, each 2.0 N or smaller. Record the spring scale readings in the table below, and then use the **analysis feature** of the software and record the corresponding force probe readings.

Spring scale reading (N)			
Force probe reading (N)			

Question 1-8: Do your force probe readings correspond to your spring scale readings? Can you now use the force probe to make reasonably accurate force measurements?

INVESTIGATION 2: MOTION AND FORCE

Now you can use the force probe to apply known forces to an object. You can also use the motion detector, as in the previous two labs, to examine the motion of the object. In this way you will be able to explore the relationship between motion and force.

You will need the following materials:

- computer-based laboratory system
- *RealTime Physics Mechanics* experiment configuration files
- force probe
- motion detector
- spring scale with a maximum reading of 5 N
- cart with very little friction
- masses to increase cart's mass to about 1 kg
- smooth ramp or other level surface 2–3 m long
- low-friction pulley, lightweight string, table clamp, variety of hanging masses

Activity 2-1: Pushing and Pulling a Cart

In this activity you will move a low friction cart by pushing and pulling it with your hand. You will measure the force, velocity, and acceleration. Then you will be able to look for mathematical relationships between the applied force and the velocity and acceleration, to see whether either is (are) related to the force.

1. Set up the cart, force probe, and motion detector on a smooth level surface as shown below. The cart should have a mass of about 1 kg with force probe included. Fasten additional mass to the top if necessary.

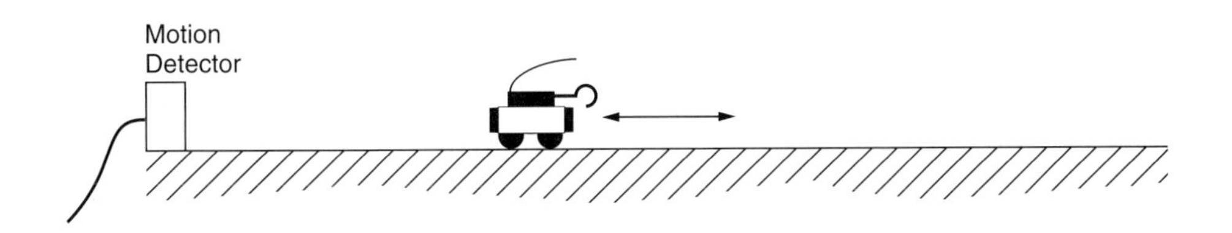

The force probe should be fastened *securely* to the cart *so that its body and cable do not extend beyond the end of the cart facing the motion detector. (Tape the force probe cable back along the body to assure that it will not be seen by the motion detector.)*

Prediction 2-1: Suppose you grasp the force probe hook and move the cart forward and backward in front of the motion detector. Do you think that either the velocity or the acceleration graph will look like the force graph? Is either of these motion quantities related to force? (That is to say, if you apply a changing force to the cart, will the velocity or acceleration change in the same way as the force?) Explain.

2. To test your predictions, open the experiment file called **Motion and Force (L03A2-1).** This will set up velocity, force, and acceleration axes with a convenient time scale of 5 s, as shown below. **Calibrate** the force probe using a 2.0-N pull from a spring scale, as in Activity 1-4, or **load the calibration,** if you haven't already done so.

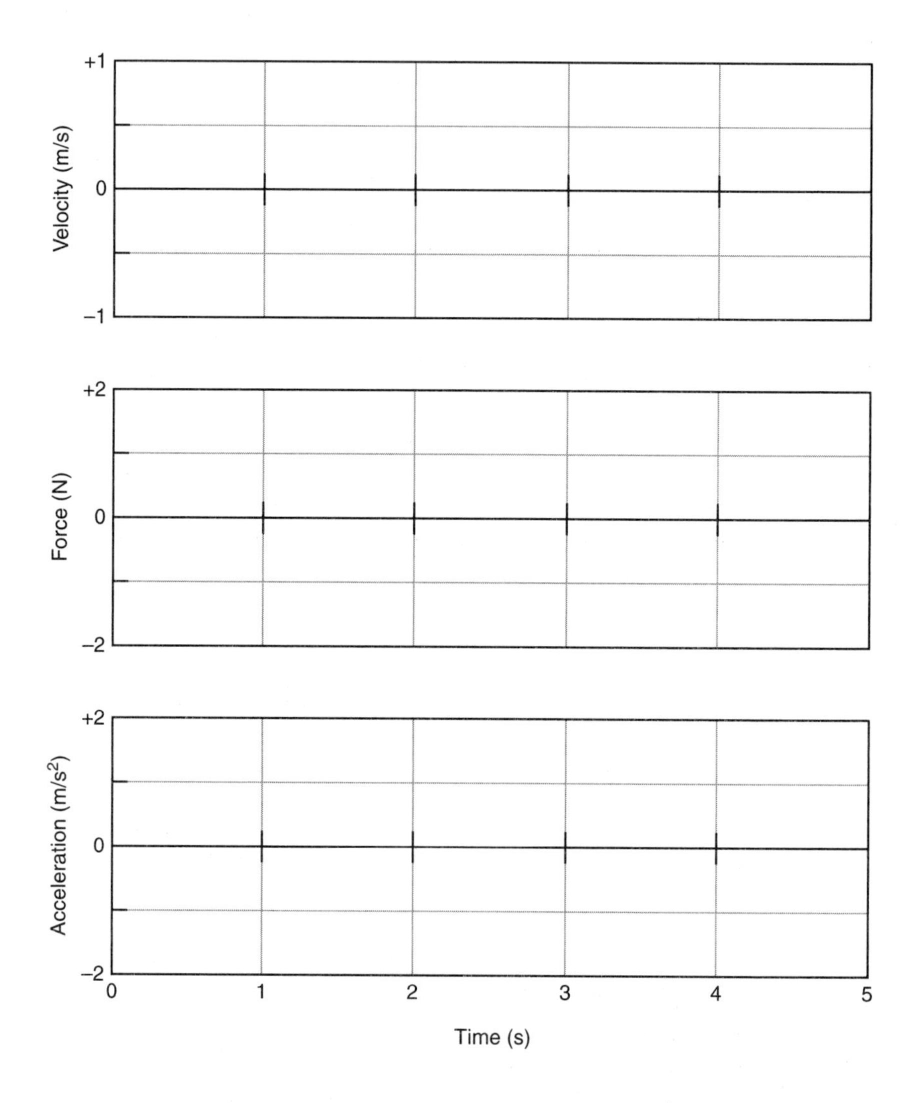

3. **Zero** the force probe. Grasp the force probe hook and **begin graphing.** When you hear the clicks, pull the cart quickly away from the motion detector and stop it quickly. Then push it quickly back toward the motion detector and again stop it quickly. Try to get sudden starts and stops, and to pull and push the force probe hook along a straight line parallel to the ramp. *Do not twist the hook. Be sure that the cart never gets too close to the motion detector. Be sure your hand and body are out of the way of the motion detector.*

4. Carefully sketch your graphs on the axes above, or **print** them and affix the graphs over the axes.

Question 2-1: Does either graph—velocity or acceleration—resemble the force graph? Which one? Explain how you reached this conclusion.

Question 2-2: Based on your observations, does it appear that there is a mathematical relationship between either applied force and velocity, applied force and acceleration, both, or neither? Explain.

Activity 2-2: Speeding Up Again

You have seen in the previous activity that force and acceleration seem to be related. But just what is the relationship between force and acceleration?

Prediction 2-2: Suppose that you have a cart with very little friction and you pull this cart with a constant force as shown below on the force–time graph. Sketch on the axes below your predictions of the velocity–time and acceleration–time graphs of the cart's motion.

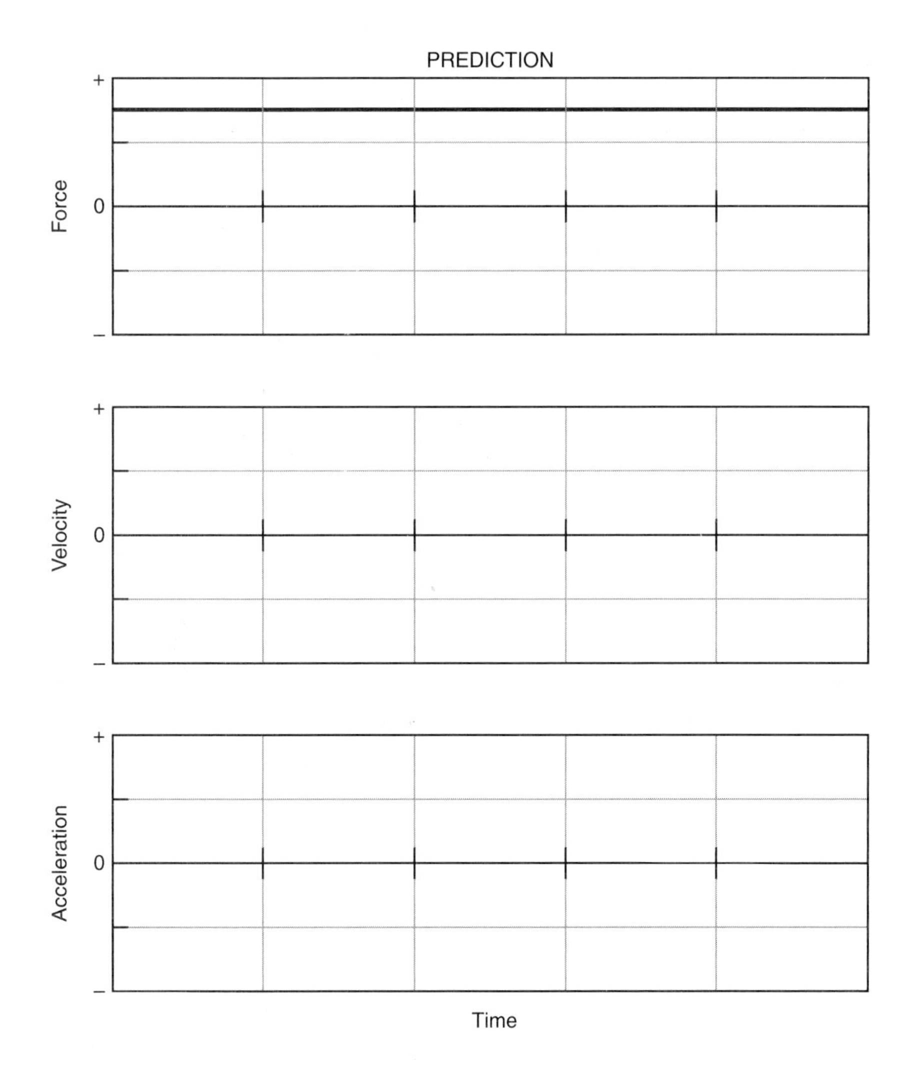

Describe in words the predicted shape of the velocity vs. time and acceleration vs. time graphs that you sketched.

1. Test your predictions. Set up the ramp, pulley, cart, string, motion detector, and force probe as shown below. The cart should be the same mass as before (about 1 kg).

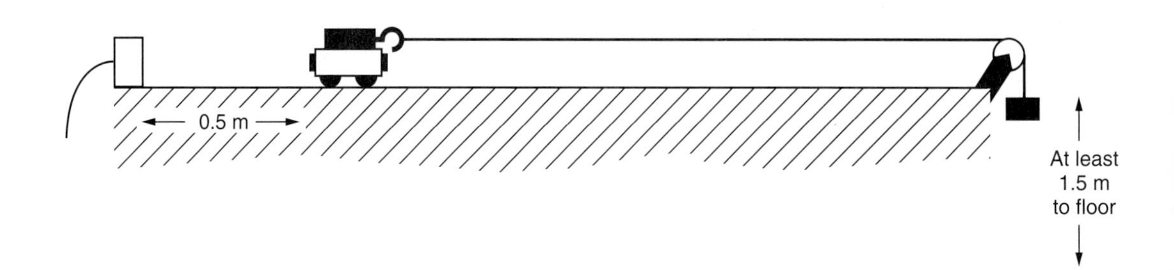

   Be sure that the cart's friction is minimum. (If the cart has a friction pad, it should be raised so it doesn't contact the ramp.)

2. Prepare to graph velocity, acceleration, and force. Open the experiment file called **Speeding Up Again (L03A2-2)** to display the velocity, acceleration, and force axes that follow.

3. It is important to choose the amount of the falling mass so the cart doesn't move too fast to observe the motion. Experiment with different hanging masses until you can get the cart to move across the ramp in about 2–3 s after the mass is released.

   Record the hanging mass that you decided to use: ______

   Also test to be sure that the motion detector sees the cart during its complete motion. *Remember that the back of the cart must always be at least 0.5 m from the motion detector.*

4. **Calibrate** the force probe with a force of 2.0 N applied to it with the spring scale, or **load the calibration** if you haven't already done so.

5. **Zero** the force probe with the string hanging loosely so that no force is applied to the probe. **Zero** it again *before each graph.*

6. **Begin graphing.** Release the cart after you hear the clicks of the motion detector. *Be sure that the cable from the force probe is not seen by the motion detector, and that it doesn't drag or pull the cart.*

   Repeat until you get good graphs in which the cart is seen by the motion detector over its whole motion.

7. Transfer your data so that the graphs will be **persistently displayed on the screen.**

8. If necessary, **adjust the axes** to display the graphs more clearly. Sketch the actual velocity, acceleration, and force graphs on the axes that follow, or **print** the graphs and affix them over the axes. Draw smooth graphs; don't worry about small bumps.

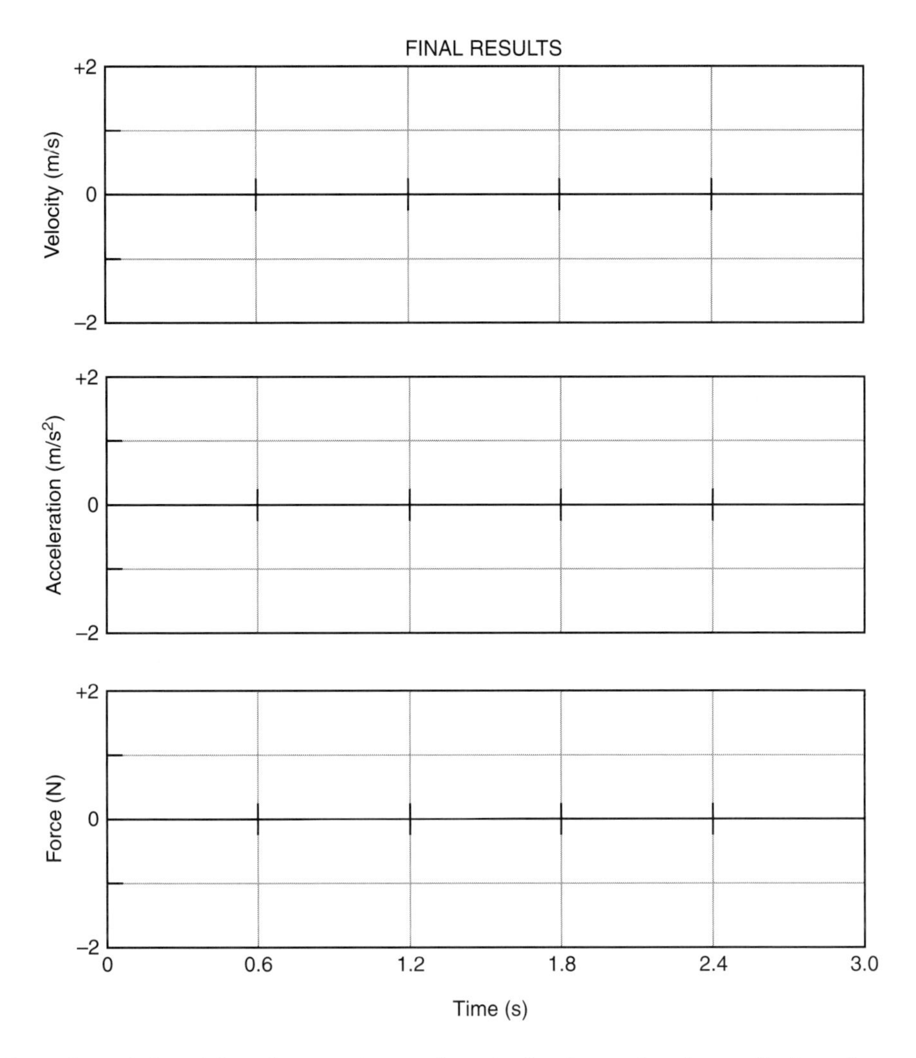

Questions 2-3: After the cart is moving, is the force that is applied to the cart by the string constant, increasing, or decreasing? Explain based on your graph.

Question 2-4: How does the acceleration graph vary in time? Does this agree with your prediction? Does a constant applied force produce a constant acceleration?

Question 2-5: How does the velocity graph vary in time? Does this agree with your prediction? What kind of change in velocity corresponds to a constant applied force?

Activity 2-3: Acceleration From Different Forces

In the previous activity you examined the motion of a cart with a constant force applied to it. But what is the relationship between acceleration and force? If you apply a larger force to the same cart (while the mass of the cart is not changed) how will the acceleration change? In this activity you will try to answer these questions by applying different forces to the cart, and measuring the corresponding accelerations.

If you accelerate the same cart with two other different forces, you will then have three data points—enough data to plot a graph of acceleration vs. force. You can then find the mathematical relationship between acceleration and force (with the mass of the cart kept constant).

Prediction 2-3: Suppose you pulled the cart with a force about twice as large as before. What would happen to the acceleration of the cart? Explain.

1. Test your prediction. Keep the graphs from Activity 2-2 **persistently displayed on the screen.**
2. Accelerate the cart with a larger force than before. To produce a larger force, hang a mass about two times as large as in the previous activity.

   Record the hanging mass: ______
3. Graph force, velocity, and acceleration as before. *Don't forget to* ***zero*** *the force probe with nothing attached to the hook right before graphing.*
4. Sketch your graphs as well as the graphs from Activity 2-2 on the axes below, or **print** out the graphs and affix them over the axes.

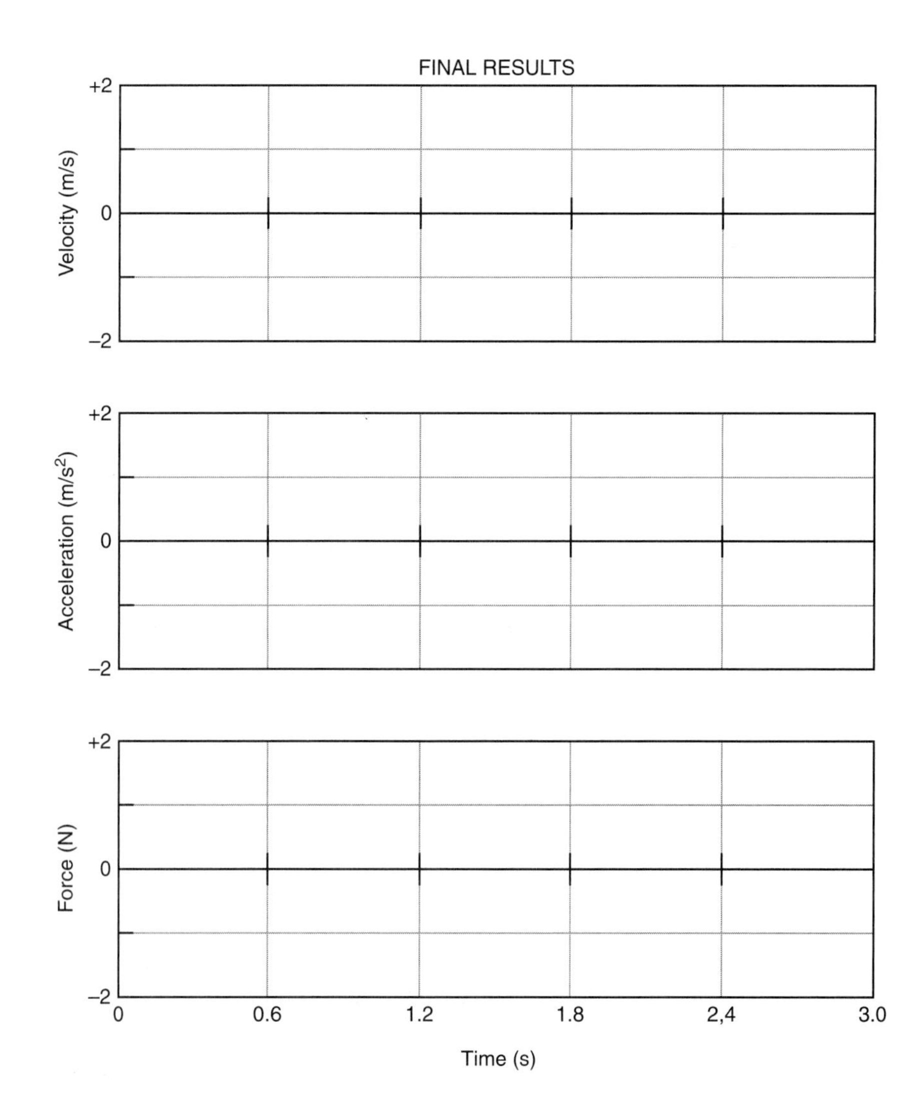

Table 2-3

	Average force (N)	Average acceleration (m/s^2)
Activity 2-2		
Activity 2-3		

5. Use the **statistics feature** of the software to measure the average force and average acceleration for the cart for this activity and Activity 2-2, and record your measured values in Table 2-3. *Find the mean values only during the time intervals when the force and acceleration are nearly constant.*

Question 2-6: How did the force applied to the cart compare to that with the smaller force in Activity 2-2?

Question 2-7: How did the acceleration of the cart compare to that caused by the smaller force in Activity 2-2? Did this agree with your prediction? Explain.

6. Accelerate the cart with a force roughly midway between the other two forces you applied. Use a hanging mass about midway between those used before.

   Record the mass: ______

7. Graph velocity, acceleration, and force. Sketch the graphs on the previous axes, or **print** and affix them.
8. Find the mean acceleration and force, as before, and record the values in the bottom row of Table 2-3.

If you have time, carry out the following Extension to get more data for the acceleration vs. force graph you will make in the next activity.

Extension 2-4: More Acceleration vs. Force Data

Gather data for average acceleration and average force for several other forces applied to the same cart. Add additional rows to Table 2-3.

Activity 2-5: Relationship Between Acceleration and Force

In this activity you will find the mathematical relationship between acceleration and force.

1. Use the **graphing routine** in the software to plot a graph of acceleration vs. force from the data in Table 2-3. To plot the graph load the experiment file

called **Acceleration vs. Force (L03A2-5),** and **enter** your data in the table. You may wish to **adjust the graph axes,** after all of the data are entered, to better display the data.

2. Use the **fit routine** to determine the mathematical relationship between the acceleration of the cart and the force applied to the cart as displayed on your graph.
3. **Print** your graph along with the fit equation and affix it in the space below.

Question 2-8: Does there appear to be a simple mathematical relationship between the acceleration of a cart (with fixed mass and negligible friction) and the force applied to the cart (measured by the force probe mounted on the cart)? Write down the equation you found and describe the mathematical relationship in words.

Question 2-9: If you increased the force applied to the cart by a factor of 10, how would you expect the acceleration to change? How would you expect the acceleration–time graph of the cart's motion to change? Explain based on your graphs.

Question 2-10: If you increased the force applied to the cart by a factor of 10, how would you expect the velocity–time graph of the cart's motion to change? Explain based on your graphs.

Comment: The mathematical relationship that you have been examining between the acceleration of the cart and the applied force is known as *Newton's second law.* In words, when there is only one force acting on an object, the force is equal to the mass of the object times its acceleration.

Name______________________ Date____________ Partners______________________

Homework for Lab 3: Force and Motion

1. You are given 10 identical springs. Describe how you would develop a scale of force (i.e., a means of producing repeatable forces of a variety of sizes) using these springs.

2. Describe how you would use an uncalibrated force probe and the springs in Question 1 to develop a quantitative scale of force. How could you measure forces that do not correspond to exact numbers of stretched springs?

3. What is meant by a proportional relationship? Is this the same as a linear relationship? Explain.

4. Given the table of data below for widgets and doodads, how would you determine if the relationship between widgets and doodads is a proportional one? Sketch on the axes to the right of the table what the graph would look like if widgets are proportional to doodads.

Widgets	Doodads
0.0	0.0
150.5	10.0
305.0	20.0
442.7	30.0
601.3	40.0

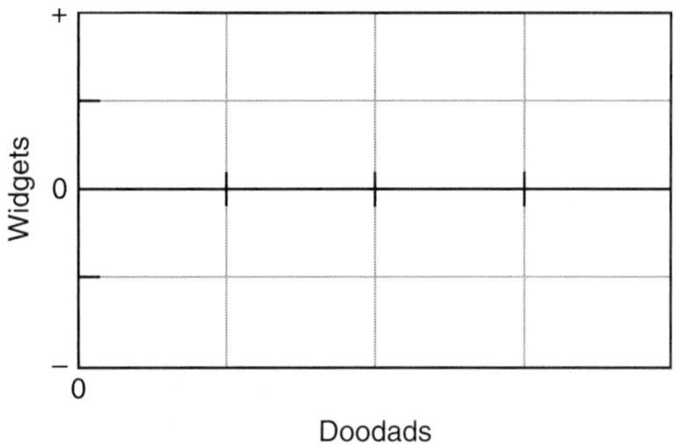

5. A force is applied that makes an object move with the acceleration shown below. Assuming that friction is negligible, sketch a force–time graph of the force on the object on the axes below.

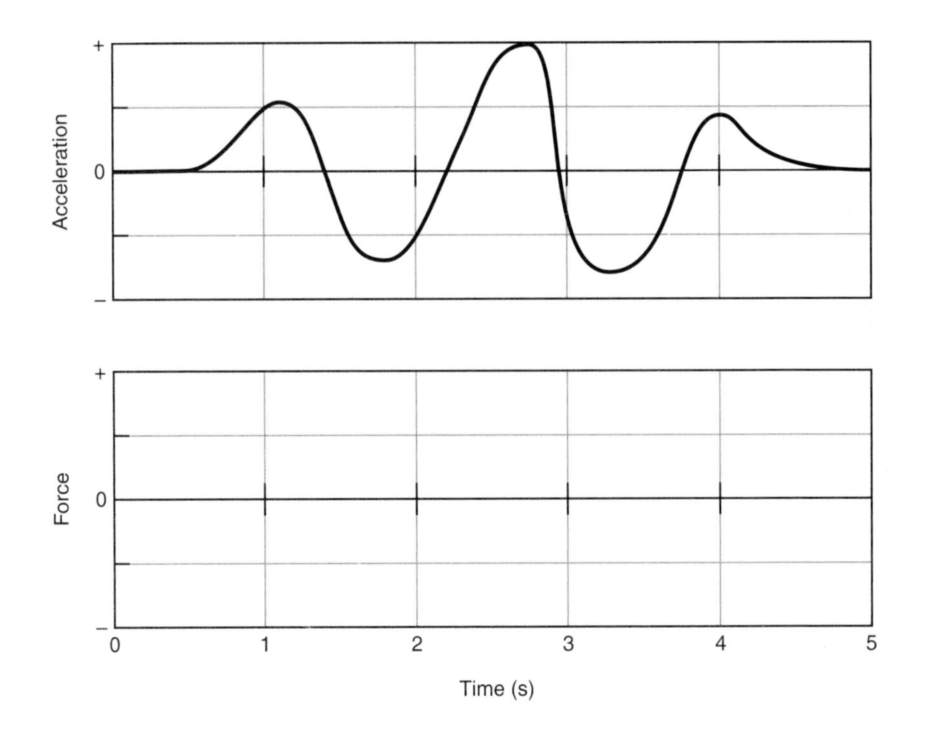

Explain your answer:

6. Roughly sketch a possible velocity–time graph for the object in Question 5 on the axes below.

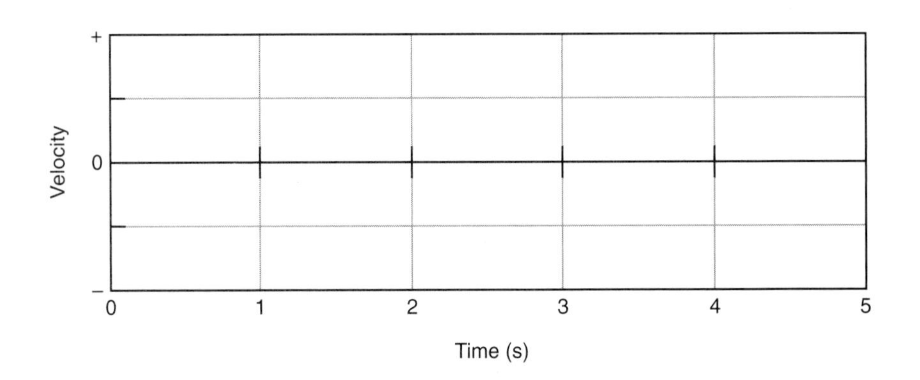

7. A cart can move along a horizontal line (the + position axis). It moves with the velocity shown below.

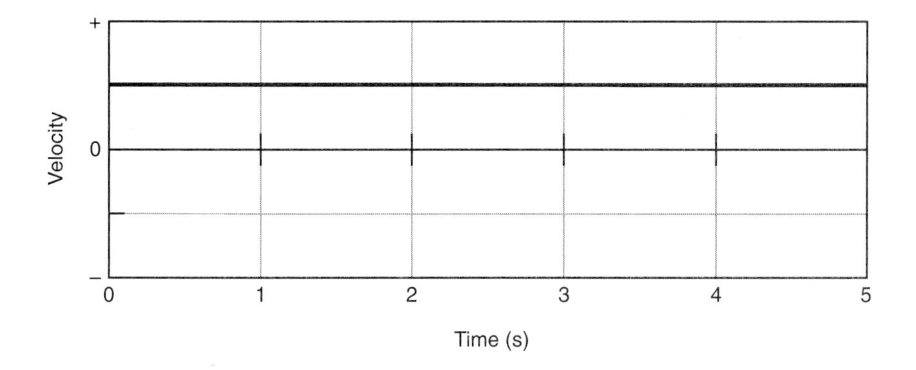

Sketch the acceleration–time graph of the cart's motion on the axes below.

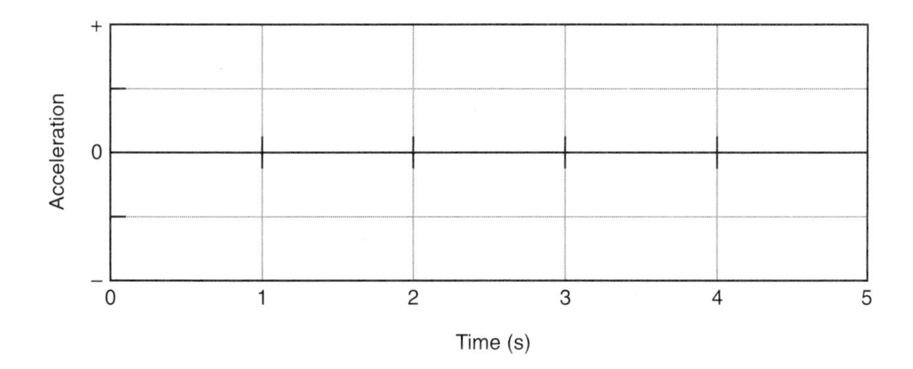

Assuming that friction is so small that it can be neglected, sketch on the axes below the force that must act on the cart to keep it moving with this velocity and acceleration.

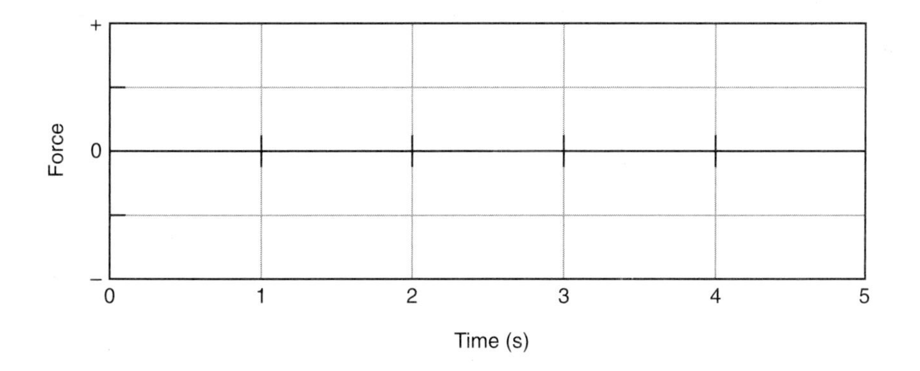

Explain both of your graphs.

Questions 8–10 refer to an object that can move in either direction along a horizontal line (the + position axis). Assume that friction is so small that it can be neglected. Sketch the shape of the graph of the force applied to the object that would produce the motion described.

8. The object moves away from the origin with a constant velocity.

9. The object moves toward the origin with a constant velocity.

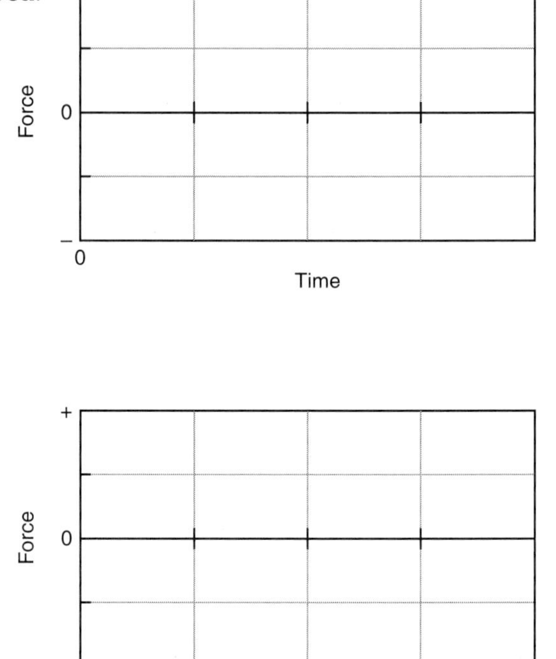

10. The object moves away from the origin with a steadily increasing velocity (a constant acceleration).

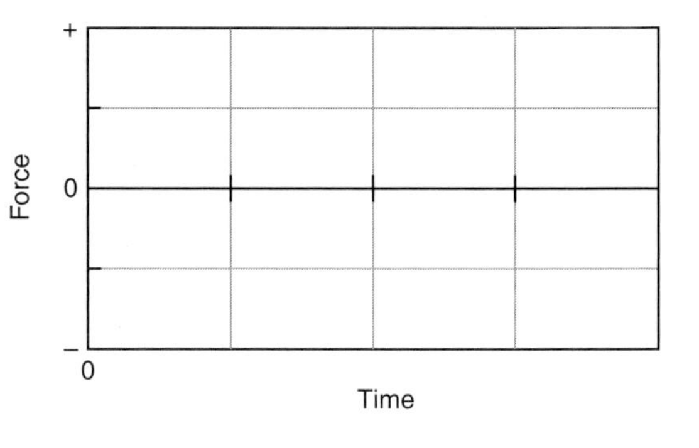

Questions 11 and 12 refer to an object that can move along a horizontal line (the + position axis). Assume that friction is so small that it can be ignored. The object's velocity–time graph is shown on the right.

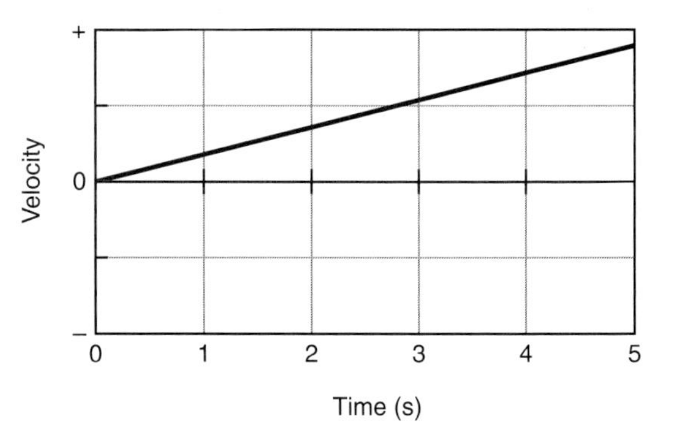

11. Sketch the shapes of the acceleration–time and force–time graphs on the axes below.

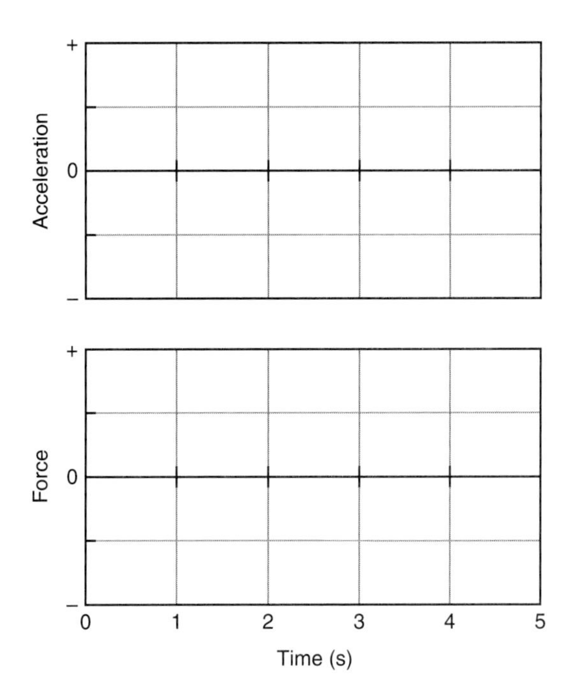

12. Suppose that the force applied to the object were twice as large. Sketch with dashed lines on the same axes above the force, acceleration, and velocity.

13. An object can move along a horizontal line (the + position axis). Assume that friction is so small that it can be ignored. The object's velocity–time graph is shown below.

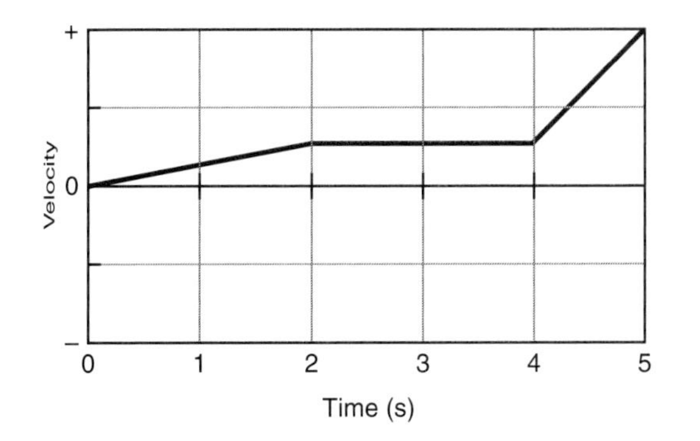

Sketch the shapes of the acceleration and force graphs on the axes below.

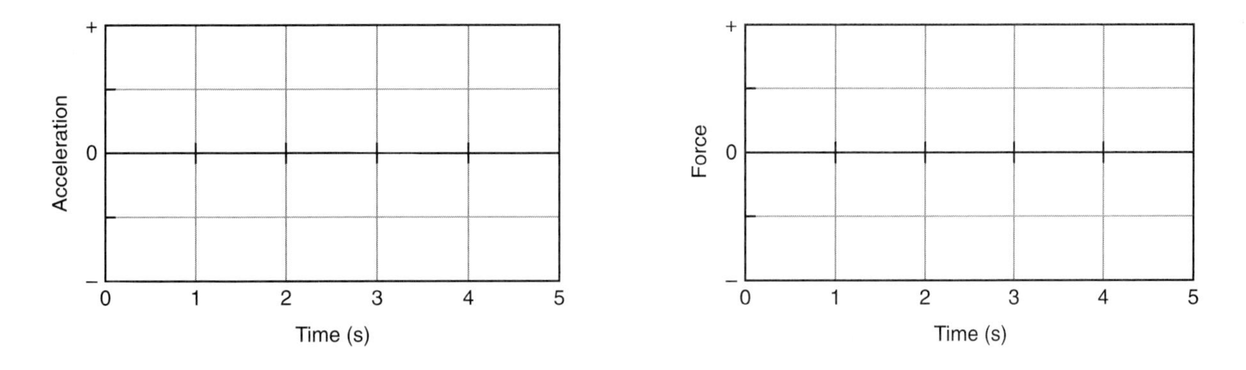

Name______________________________ Date________________

Pre-Lab Preparation Sheet for Lab 4: Combining Forces

(Due at the beginning of Lab 4)

Directions:
Read over Lab 4 and then answer the following questions about the procedures.

1. A cart is moving toward the right and slowing down, as in Activity 1-1. Predict the direction of the combined (net) force on the cart.

2. In Activity 2-1, what are the spring scales used for?

3. Why are two fan units used in Activity 2-2?

4. Sketch your Prediction 3-2 on the axes below.

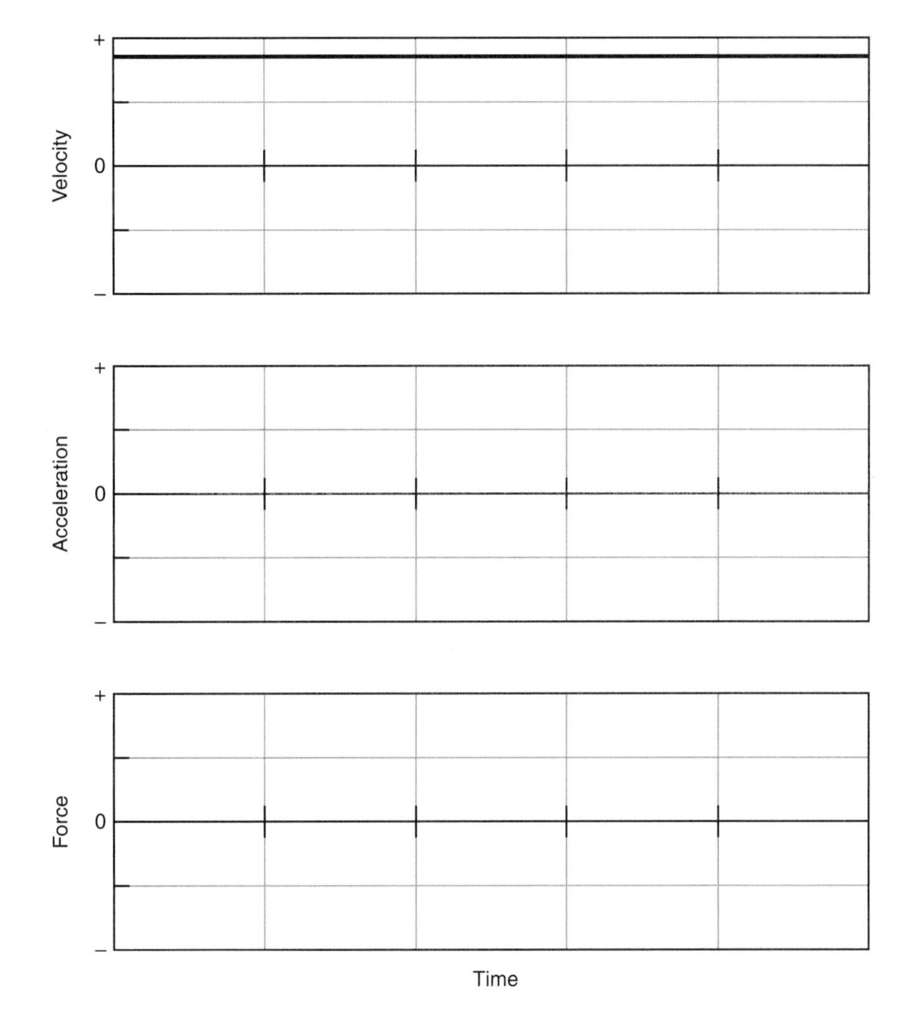

Name______________________________ Date____________ Partners________________________________

Lab 4: Combining Forces

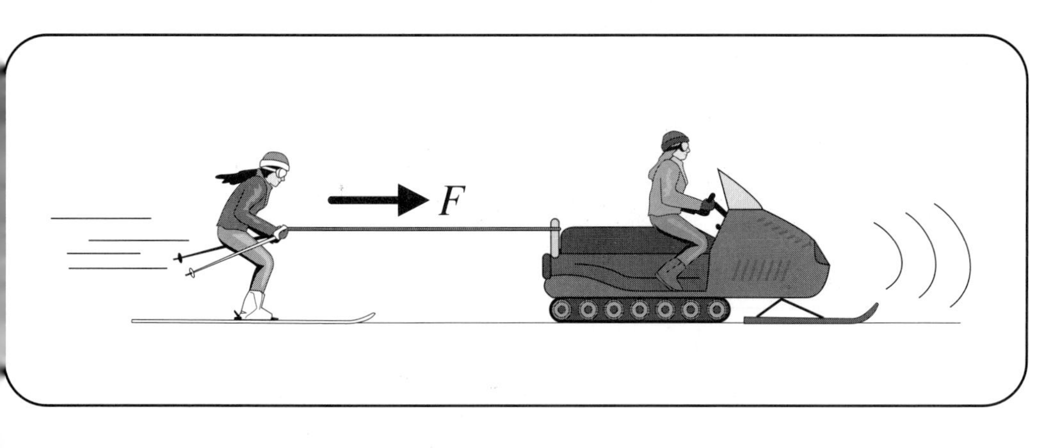

Nature and Nature's laws lay hid at night:
God said, "Let Newton be!" and all was light.

—Alexander Pope

OBJECTIVES

- To understand the relationship between the direction of the force applied to an object and the direction of the acceleration of the object.
- To understand how different forces can act together to make up a *combined force.*
- To establish a definition of *combined force* as that which changes an object's motion according to *Newton's second law.*
- To understand the motion of an object with no net force applied to it and how *Newton's first law* describes this motion.

OVERVIEW

In previous labs you have examined the one-dimensional motions of an object caused by a single force applied to the object. You have seen that when friction is so small that it can be ignored, a single constant applied force will cause an object to have a constant acceleration. (The object will speed up at a steady rate.)

Under these conditions, you have seen that the acceleration is proportional to the applied force, if the mass of the object is not changed. You saw that when a constant force is applied to a cart with very low friction, the cart speeds up at a constant rate so that it has a constant acceleration. If the applied force is made larger, then the acceleration is proportionally larger. This allows you to define force more precisely not just in terms of the stretches of rubber bands and springs, but as the *"thing" that causes acceleration.*

The major goal of this lab is to continue to develop the relationship between force and acceleration: the first two of Newton's famous laws of motion.

In Investigation 1, you will explore motions in which the applied force (and hence the acceleration of the object) is in a different direction than the object's velocity. In such a case the force will cause the object to slow down. (Its speed decreases.)

In Investigation 2, you will explore what happens when more than one force is applied to an object.

In Investigation 3, you will be asked to consider the special case when the object moves with a constant velocity, so that the object's acceleration is zero. What combination of forces must be applied to an object to keep it moving with a constant velocity when there is almost no friction? You can answer this question by collecting force and motion data again with the force probe and motion detector. This will lead you to the discovery of *Newton's first law* (in a situation where friction is so small that it can be neglected).

INVESTIGATION 1: SPEEDING UP AND SLOWING DOWN

In Lab 3 you looked at cases where the velocity, force, and acceleration *all have the same sign* and are all positive. That is, the vectors representing each of these three vector quantities *all point in the same direction.* For example, if the cart is moving toward the right and a force is exerted toward the right, then the cart will speed up. The acceleration is also toward the right. The three vectors can be represented as

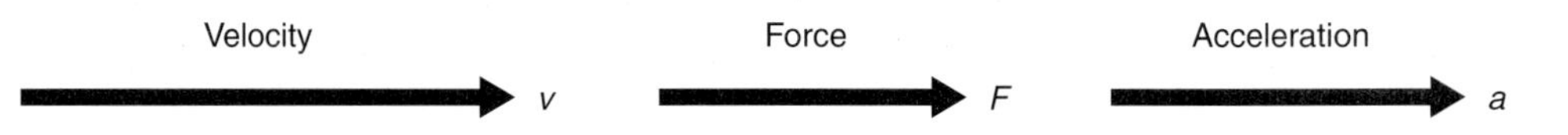

If the positive x direction is toward the right, then you could also say that the velocity, acceleration, and force are all *positive.* In this investigation, you will examine the vectors representing velocity, force, and acceleration for other motions of the cart. This will be an extension of your observations in Lab 2, Changing Motion.

You will need the following:

- computer-based laboratory system
- *RealTime Physics Mechanics* experiment configuration files
- force probe
- motion detector
- low-friction cart
- spring scale with a maximum reading of 5 N
- smooth ramp or other level surface 2–3 m long
- low-friction pulley and string
- variety of small hanging masses (10–50 g)

Activity 1-1: Slowing Down Away from the Motion Detector

1. Set up the cart, ramp, pulley, hanging mass, and motion detector as shown in the diagram that follows. You may need to position the motion detector slightly off to the side of the pulley so that it "sees" the cart and not the string.

Now when you give the cart a push away from the motion detector, it will slow down after it is released. In this activity you will examine the acceleration and the applied force.

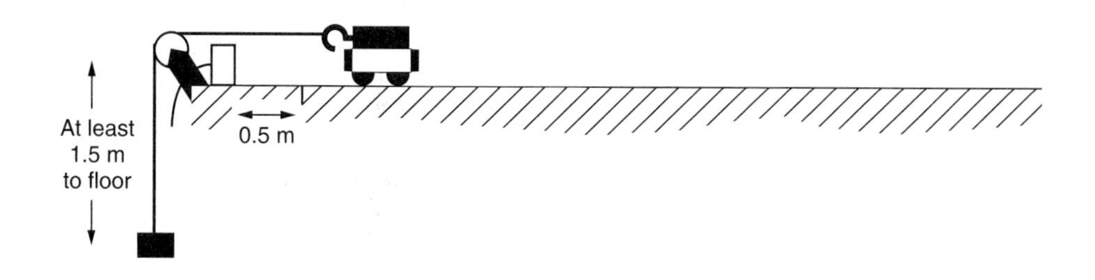

Prediction 1-1: Suppose that you give the cart a push toward the right and release it. Draw below vectors that might represent the velocity, force, and acceleration of the cart at each time after it is released and is moving toward the right. Be sure to mark your arrows with v, F, or a as appropriate. Assume that the cart is moving at t_1 and t_4.

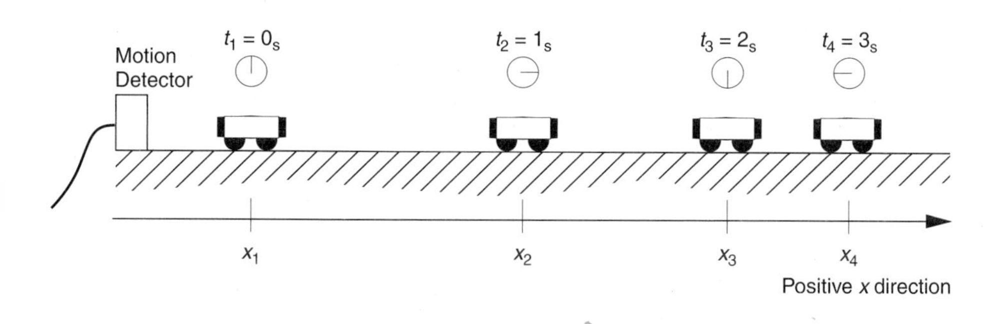

If the positive x direction is toward the right, what are the signs of the velocity, force, and acceleration after the cart is released and is moving toward the right?

Velocity Force Acceleration

2. Test your predictions. Use a hanging mass that causes the cart to move all the way across the ramp from right to left in about 2–3 s when the mass is released.

   Record the hanging mass that you decided to use:______

3. Open the experiment file called **Slowing Down Again (L04A1-1)** to display the axes that follow.

4. Test to be sure that the motion detector sees the cart during its complete motion, and that the string and force probe cable are not interfering with the motion detector or the motion of the cart. *Remember that the cart must always be at least 0.5 m from the motion detector.*

5. **Calibrate** the force probe with a force of 2.0 N applied to it with the spring scale or **load the calibration.** (If you are using a Hall effect force probe, you may need to adjust the spacing and **check the sensitivity** to see that it is appropriate.)

6. As usual, the positive x direction is chosen to be away from the motion detector—toward the right. Since a push on the force probe now is a force toward the right, and is therefore positive, the software has been set up to **make a push positive** (and a pull negative).

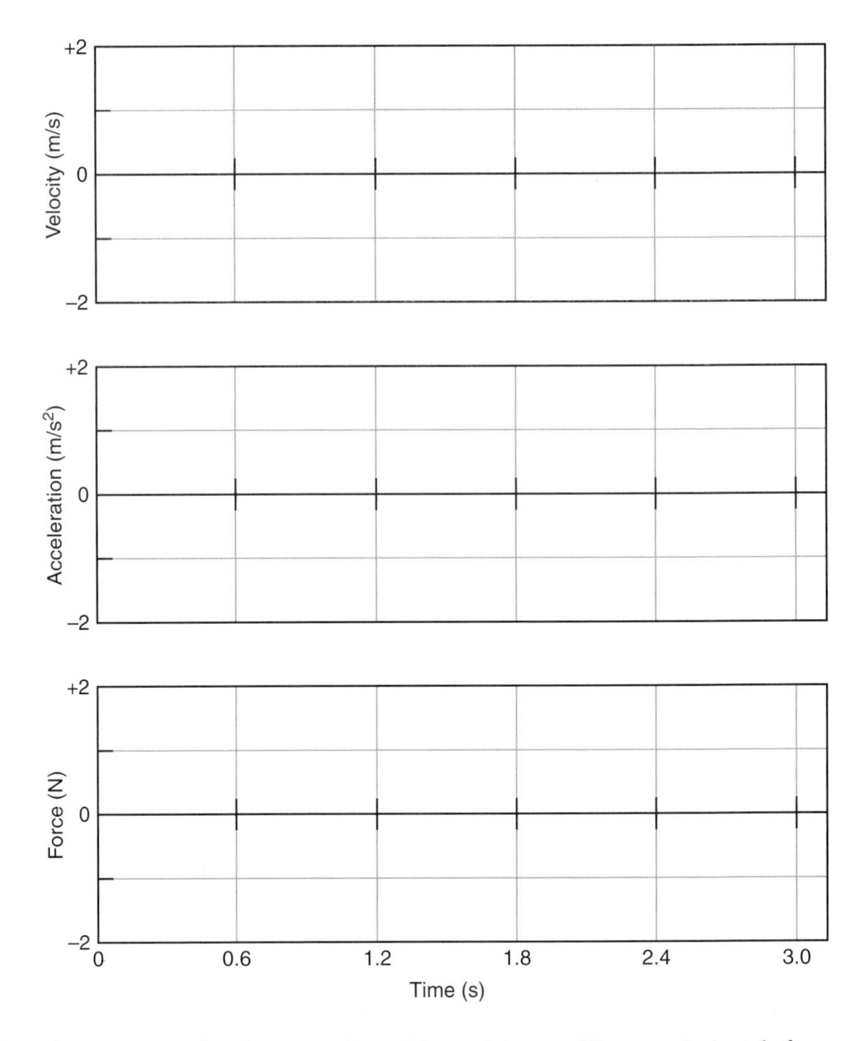

7. As always, **zero** the force probe with nothing pulling on it *just before* each graph. **Begin graphing,** and when the motion detector starts clicking, give the cart a short push toward the right and then let it go. *Be sure to keep your hand out of the region between the motion detector and the cart.* Stop the cart before it reverses its direction.

   After you have collected good graphs, move your data so that the graphs are **persistently displayed on the screen** for later comparison.

8. Sketch your velocity, acceleration, and force graphs on the axes above, or **print** them and affix them over the axes.

Question 1-1: Did the signs of the velocity, force, and acceleration agree with your predictions? If not, can you now explain the signs?

Question 1-2: Did the velocity and acceleration both have the same sign? Explain these signs based on the relationship between acceleration and velocity.

Question 1-3: Did the force and acceleration have the same sign? Were the force and acceleration in the same direction? Explain.

Question 1-4: Based on your observations, draw below vectors that might represent the velocity, force, and acceleration for the cart at the same instant in time.

Velocity Force Acceleration

Do these agree with your predictions? If not, can you now explain the directions of the vectors?

Question 1-5: After you released the cart, was the force applied by the hanging mass constant, increasing, or decreasing? Explain why this kind of force is necessary to cause the observed motion of the cart.

Activity 1-2: Speeding Up Toward the Motion Detector

Using the same setup as in the last activity, you can start with the cart at the *right* end of the ramp and release it from rest. It will then be accelerated toward the motion detector as a result of the force applied by the falling mass.

Prediction 1-2: Suppose that you release the cart from rest and let it move toward the motion detector. Draw on the diagram below vectors that might represent the velocity, force, and acceleration of the cart at each time after it is released and is moving toward the left. Be sure to mark your arrows with v, F, or a as appropriate. Assume that the cart is already moving at t_1.

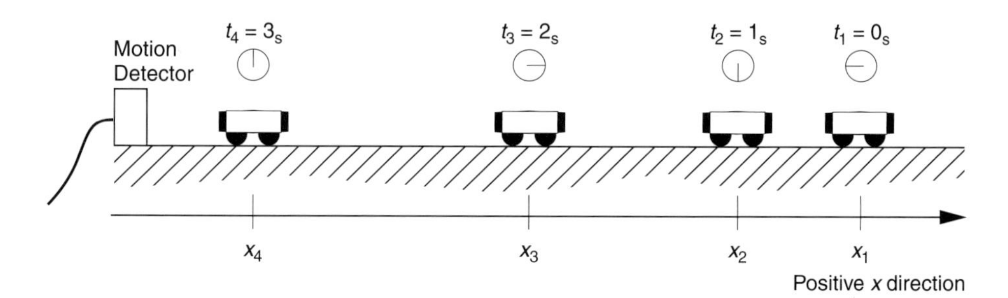

What are the signs of the velocity, force, and acceleration after the cart is released and is moving toward the motion detector? (The positive x direction is toward the right.)

Velocity Force Acceleration

1. Test your predictions. Use the same hanging mass as before. Don't forget to **zero** the force probe, with nothing pulling on it. **Begin graphing.** When you hear the motion detector, release the cart from rest as close to the right end of the ramp as possible. *Catch the cart before it hits the motion detector.*

2. Sketch your graphs on the previous axes with dashed lines, or **print** and affix them over the axes.

Question 1-6: Which of the signs—velocity, force, and/or acceleration—are the same as in the previous activity where the cart was slowing down and moving away, and which are different? Explain any differences in terms of the differences in the motion of the cart.

Question 1-7: Based on your observations, draw below vectors that might represent the velocity, force, and acceleration for the cart at the same instant in time.

Velocity Force Acceleration

Do these agree with your predictions? If not, can you now explain the directions of the vectors?

Question 1-8: Write down a simple rule that describes the relationship between the direction of the applied force and the direction of the acceleration for any motion of the cart.

Question 1-9: Is the direction of the velocity always the same as the direction of the force? Is the direction of the acceleration always the same as the direction of the force? In terms of its magnitude and direction, what is the effect of a force on the motion of an object?

If you have more time, carry out the following Extension to explore more carefully the signs of velocity, acceleration, and force.

Extension 1-3: Reversing Direction

Prediction E1-3: Suppose that you use the same setup as in Activities 1-1 and 1-2, but now you give the cart a push away from the motion detector, release it, and let it move away, reverse direction, and head back toward the motion detector. Sketch on the axes that follow with dashed lines your predictions for the velocity, acceleration, and force after the cart leaves your hand and before you stop it. Mark on your prediction the time at which the cart reverses direction.

Also describe the velocity, acceleration, and force in words in the space below.

Carry out this observation. You may use the same experiment file as in Activity 1-1, **Slowing Down Again (L04A1-1).** Sketch your observed graphs with solid lines on the axes that follow, or **print** the graphs and affix them over the axes. Mark on your graphs the time at which the cart reverses direction.

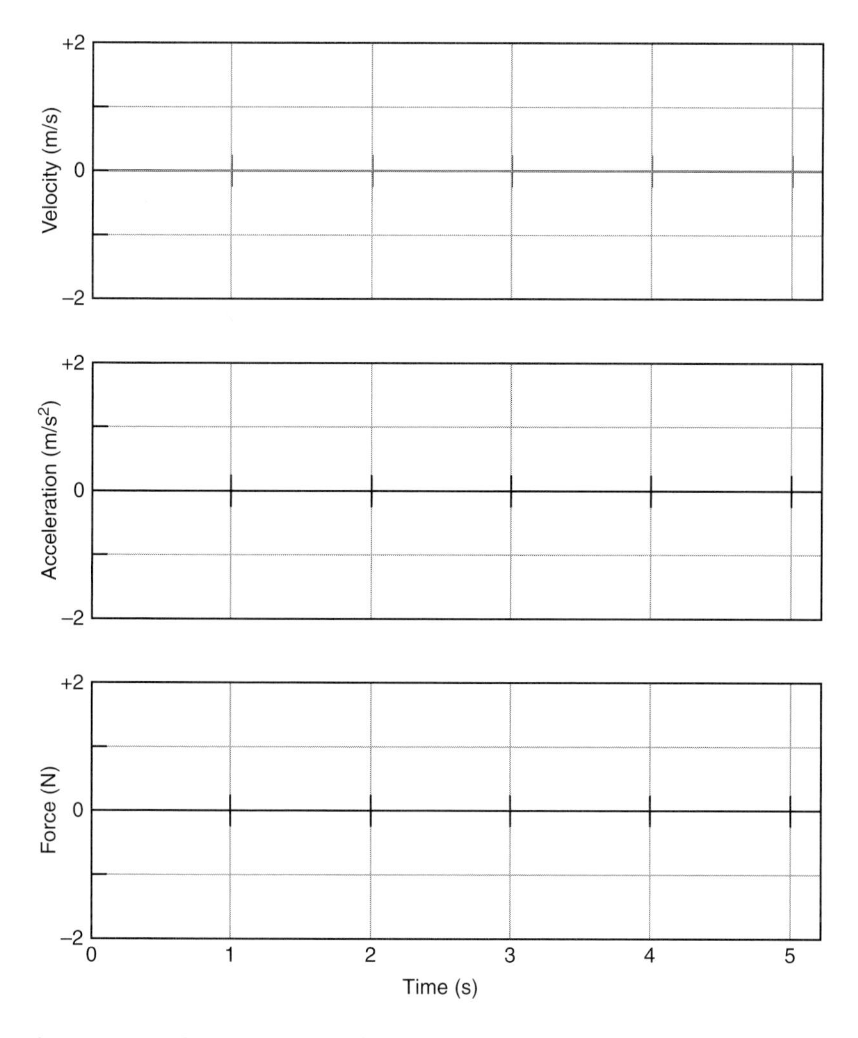

Question E1-10: Compare your observations with your predictions. Describe the force and acceleration at the moment when the cart reverses direction.

Question E1-11: Based on your knowledge of acceleration and force, explain why the force and acceleration have the signs they have at the moment the cart reverses direction.

INVESTIGATION 2: NET FORCE: COMBINING APPLIED FORCES

As you know, *vectors* are mathematical entities that have both magnitude and direction. Thus, a one-dimensional vector can point either in the positive or negative x direction. Vectors pointing in the same direction add together and vectors pointing in opposite directions subtract from each other. Quantities that have vector behavior are often denoted by a letter with a little arrow above it ($\vec{\mathbf{A}}$). The sum of several vectors is often denoted by placing a summation sign in front of a vector symbol ($\Sigma\vec{\mathbf{A}}$).

It is obvious that forces have both magnitude (i.e., strength) *and* direction. You can do some simple observations to determine whether or not one-dimensional forces actually behave like vectors. To do this you will need:

- 2 spring scales with a maximum reading of 5 N
- low-friction cart

Comment: In the diagrams below we assume that forces behave like mathematical vectors and thus, can be represented by the symbols $\vec{F}_A$, $\vec{F}_B$, and $\vec{F}^{net}$.

Activity 2-1: Do One-Dimensional Forces Behave Like Vectors?

1. Observe what happens when you hook a spring scale to one end of the cart and extend it in a horizontal direction so that its force is equal to 1.0 N in magnitude. Be sure to keep the spring scale extended to 1.0 N during the entire motion of the cart. (This is a casual observation—no need to take any data.)

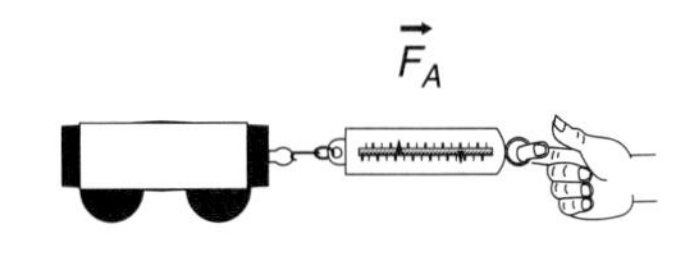

Question 2-1: Does the cart move? If so, how? Does it move with a constant velocity or does it accelerate?

Question 2-2: Draw an arrow next to the diagram above that represents a scale drawing of the magnitude and direction of the force you are applying. Let 5.0 cm of arrow length represent each newton of force. Label the arrow with an $\vec{F}_A$.

2. Examine what kind of motion results when two identical spring scales are displaced by the same amount in the same direction (for example, when each spring is displaced to give 0.5 N of force). Again keep the springs stretched throughout the entire motion.

   Compare this motion to that when one spring scale is displaced by twice that amount (for example so that it can apply 1.0 N of force as in (1) above).

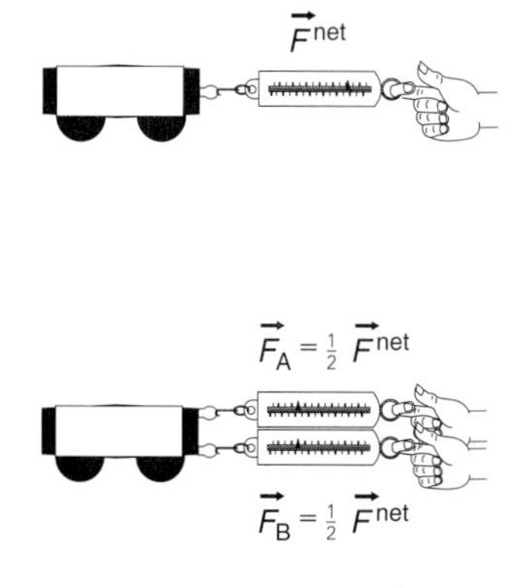

Question 2-3: Describe what you did, and compare the motions of the cart.

Question 2-4: Draw arrows next to the diagram above that represent a scale drawing of the magnitudes and directions of $\vec{F}_A$, $\vec{F}_B$, and $\vec{F}^{net}$. Again let 5.0 cm of arrow length represent each newton of force.

3. Observe what kind of motion results when two spring scales are hooked to opposite ends of the cart and extended in a horizontal direction so that *each* of their forces is equal to 1.0 N in magnitude, but they are opposite in direction.

Question 2-5: Does the cart move? If so, how? What do you think the *combined* or *net* applied force on the cart is equal to in this situation?

Question 2-6: Draw arrows next to the diagram above that represent a scale drawing of the magnitudes and directions of the forces you are applying. Let 5.0 cm of arrow length represent each newton of force. Label each arrow appropriately with an $\vec{F}_A$ or an $\vec{F}_B$.

Question 2-7: Do one-dimensional forces seem to behave like one-dimensional vectors? Why or why not?

Pulling carts with spring scales is awkward and takes practice to do well. In the rest of this investigation you will use fan units to apply forces to the cart, and the motion detector to measure the resulting velocities and accelerations of the cart.

To carry out the following activities you will need

- computer-based laboratory system
- *RealTime Physics Mechanics* experiment configuration files
- force probe
- cart with very little friction
- smooth ramp or other level surface 2–3 m long
- two identical fan units with batteries
- table clamp, rod, and rod clamp
- spring scale with a maximum reading of 5 N

Activity 2-2: Cooperating Fan Units

1. Attach both fan units to the cart as shown in the diagram that follows.

> **Note:** You want to have both fans pushing the cart toward the right (air flow toward the left). If the right-hand fan unit (fan unit B) does not have a reversing switch, either (a) put all of the batteries into this unit with the polarity reversed from normal or (b) mount this fan unit in the opposite direction from fan unit A.

In the next part you will observe the motion of the cart by placing the motion detector on the left end of the ramp.

Be sure that the fan blade of the fan unit that will be facing the motion detector does not extend beyond the front end of the cart. (If it does, the motion detector may collect bad data from the rotating blade.)

(Be sure to tape the fan units to the cart so they can't fly off.)

2. Set up the force probe in the rod clamp as shown.
3. **Calibrate** the force probe with a 2.0-N pull from a spring scale or **load the calibration.** (If you are using a magnetic (Hall effect) type force probe, be sure to **check the sensitivity** with a 2.0-N force applied, and reset the spacing if necessary.)

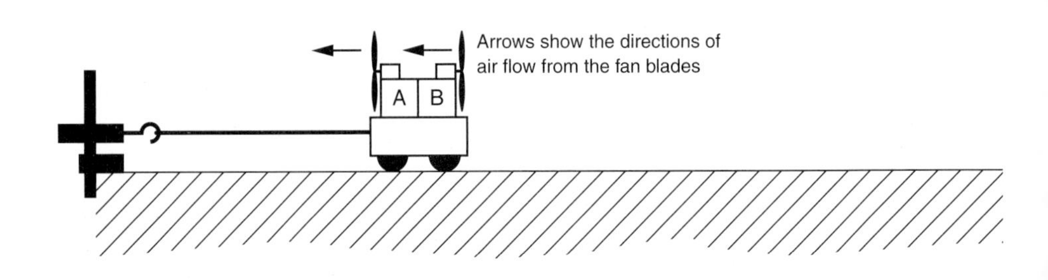

4. Use the force probe to measure the average force applied by fan unit A alone pushing the cart to the right. You may use the experiment file called **Cooperating Fan Units (L04A2-2a)** to display appropriate axes. Be sure that the software is set so that a pull on the force probe is a positive force. *To save the*

Table 2-2

	Average force (N)	Average acceleration (m/s^2)	Force/acceleration
Fan unit A alone			
Fan unit B alone			
Both fan units, A and B			

batteries, be sure to keep the fan unit turned on only during the time when you are actually making measurements.

To find the average force, use the **analysis** and **statistics features** in the software. Record your force results in Table 2-2. Leave the acceleration column blank for now.

Question 2-8: Using a scale of 10.0 cm to represent each newton of force, draw an arrow on the diagram below to represent the force vector for fan unit A.

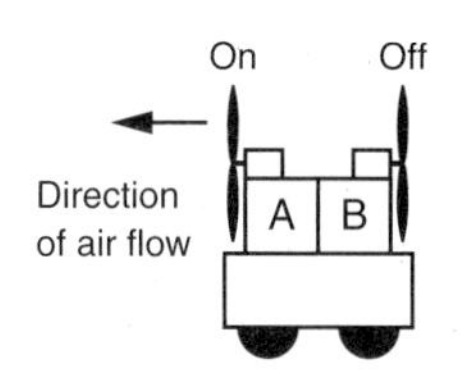

5. Place the motion detector on the left side of the ramp just in front of where the force probe is mounted. Remove the string from the cart. Open the experiment file called **Velocity and Acceleration (L04A2-2b)** to display the axes that follow.

6. Be sure that the cart starts out at least 0.5 m from the motion detector. **Begin graphing** and make acceleration and velocity graphs of the motion of the cart with this force (with just fan unit 1 turned on).

   Move your data so that the graphs are **persistently displayed on the screen** for later comparison.

Question 2-9: Does the cart move? If so, how? Does it move with a constant velocity or does it accelerate? What would you say is the effect of the single force from fan unit A toward the right on the motion of the cart?

7. Use the force probe to measure the average force applied by fan unit B alone *pushing the cart to the right.*

   Record your result in Table 2-2.

Question 2-10: Using a scale of 10.0 cm to represent each newton of force, draw an arrow on the diagram below to represent the force vector for fan unit B.

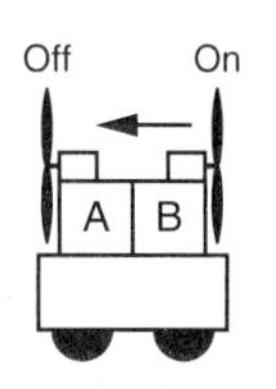

8. Place the motion detector on the left side of the ramp again, and make acceleration and velocity graphs of the motion of the cart with this force (with just fan unit B turned on).

   Sketch these graphs as well as the ones for fan unit A alone (which are persistently displayed) on the axes below, or **print** and affix them over the axes.

9. Use the **analysis** and **statistics features** of the software to find the average accelerations for fan unit A alone pushing the cart and for fan unit B alone pushing the cart, as you did for the forces. Use only the portions of the acceleration graphs where the accelerations are nearly constant.

   Record these values in Table 2-2.

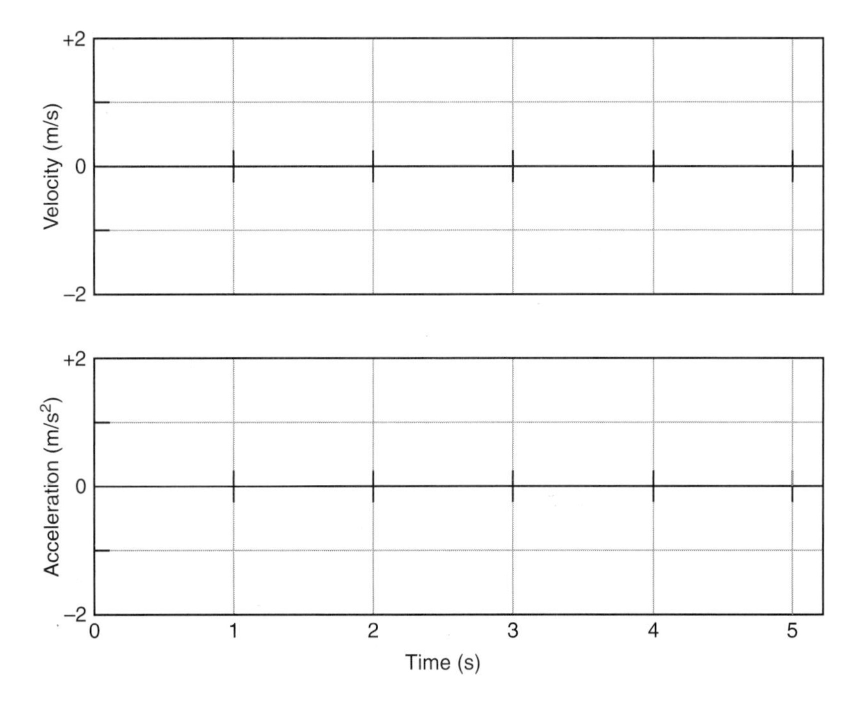

Prediction 2-1: Suppose that you turned on both fan units at the same time. What do you suppose would be the effect of the combination of the two forces on the motion of the cart? In your prediction, compare the velocity and acceleration graphs you expect with those you made above.

10. Test your prediction. Set up the motion detector and graph velocity and acceleration with both fan units pushing the cart toward the right. Determine the average acceleration as before, and record it in Table 2-2.

Question 2-11: Was your prediction correct? How does the motion of the cart with the combination of forces compare with the motion with the single force? Is the acceleration constant? Do one-dimensional forces seem to behave like one-dimensional vectors? Why or why not?

If forces are vectors, we can denote combinations of forces, or net force, such as those applied to the cart above, as

$$\vec{F}^{\text{net}} = \Sigma\vec{F} = \vec{F}_{\text{A}} + \vec{F}_{\text{B}}$$

where $\Sigma\vec{F}$ represents the vector sum of two or more forces. Some textbooks refer to a *combination of* forces or a *combined* force as a *net* force. Other authors write about the *resultant* force. *Combined, resultant,* or *net* force all refer to the same thing.

You should note that velocities and accelerations are also vectors. They also have magnitudes *and* directions.

Prediction 2-2: You measured the force applied by fan unit A and that applied by fan unit B separately above. If forces add as vectors, what value would you expect to measure for the magnitude and direction of the combination of forces applied by both fan units on the cart?

11. Use the force probe to measure the force with both fan units turned on. Find the average force, and record this value in Table 2-2.

Question 2-12: Using a scale of 10.0 cm to represent each newton of force, draw an arrow on the diagram below to represent the combined force vector.

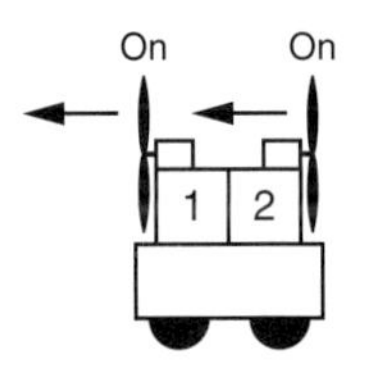

Question 2-13: Is your measured value about the same as you expected? Do one-dimensional forces seem to behave like one-dimensional vectors? Explain.

12. Calculate the ratios of the forces in Table 2-2 to the average accelerations, and record them in the last column.

Question 2-14: How do they compare? Can you explain why? When two forces act on the cart at the same time, does the acceleration appear to be proportional to the *combined* force?

Comment: In this investigation you have examined the effects of combining the forces applied by fan units, looking at the motion of a cart with either of two fan units pushing alone, and with both pushing together in the same direction. You first tested the idea that forces can be combined by the same rules as one-dimensional vectors. Then you saw that the acceleration of an object depends on the *combined* force acting on it. In other words, it is the combined force that should be used in *Newton's second law* to determine the acceleration.

In the next investigation you will carry this investigation further and look at the motion of the cart when the two fan units push in opposite directions, i.e., when the forces on the cart combine to cancel each other.

INVESTIGATION 3: MOTION AT A CONSTANT VELOCITY

In the previous investigation, you used the force probe to measure two individual forces applied to a cart in the same direction and also the combination of the two forces. Then you used the motion detector to investigate the motion of the cart. You saw that it is the *combined* force (the vector sum of the two forces) that determines the acceleration of the cart.

In this investigation you will examine what happens when the two forces applied to the cart push in opposite directions. What is the acceleration then? You will also answer the question, what *combined* force will keep the cart moving at a *constant* velocity?

You will need the following materials:

- computer-based laboratory system
- *RealTime Physics Mechanics* experiment configuration files
- motion detector
- low-friction cart
- smooth ramp or other level surface
- 2 identical fan units

Activity 3-1: Dueling Fan Units

1. If you reversed the polarity of the batteries in fan unit B for Activity 2-2, change them back to the original polarity so that the two fan units will now push the cart in opposite directions.
2. Set up the ramp, cart with two fan units, and motion detector as in Investigation 2. *Be sure that the ramp is level. Also be sure that if the fan blade is facing the motion detector the blade does not extend beyond the front end of the cart.*

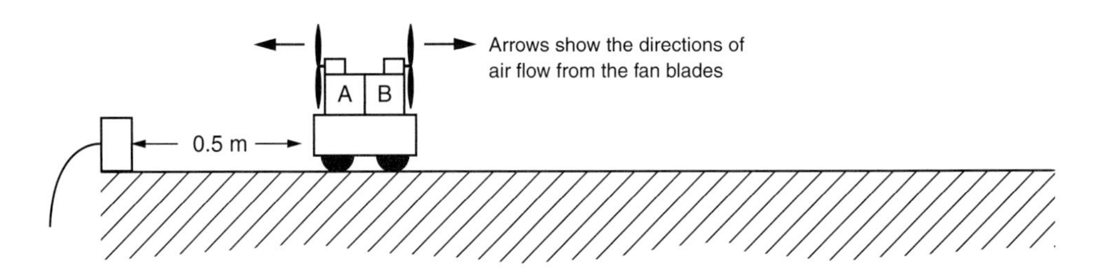

Prediction 3-1: Suppose that each fan unit is pushing on the cart with the same magnitude force. What do you predict the combined force acting on the cart will be? What do you predict the acceleration of the cart will be? Given your prediction for the acceleration, how do you predict the velocity will change?

Prediction 3-2: The cart is moving with the velocity shown on the velocity–time graph that follows. Sketch on the axes the acceleration–time graph of the cart and the force–time graph of the combined force *after the cart begins moving.*

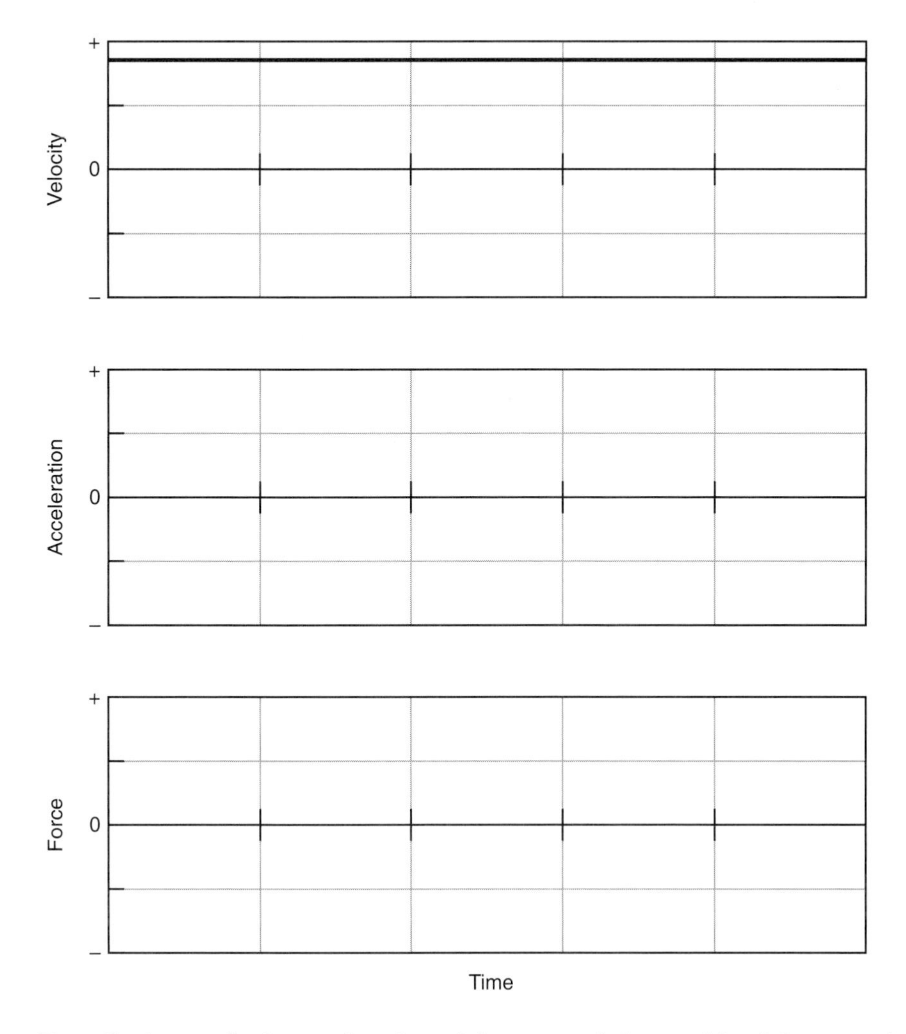

Describe in words the acceleration of the cart and the combined force needed to keep it moving at a *constant* velocity.

Prediction 3-3: Draw arrows above the carts shown in the following diagram representing the magnitude and direction of the combined force you think is needed on the cart at $t = 0$ s, $t = 1$ s, etc., to maintain its motion at a constant velocity. Assume that the cart is already moving at t_1.

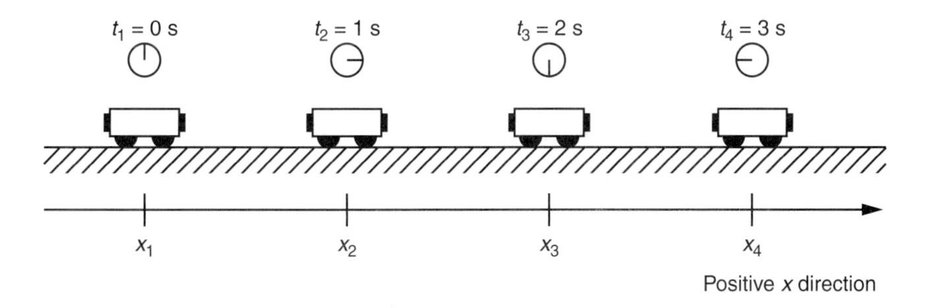

Explain the reasons for your predictions.

3. Test your predictions. Turn on both fan units. If the cart tends to move in one direction or the other, exchange batteries between the fan units (or turn the speed adjustment control) until the cart remains at rest with both fan units turned on.

Question 3-1: When both fan units are turned on, but the cart has no tendency to move, what is the *combined* (net) force exerted by both fan units on the cart? How do you know?

4. Prepare to graph velocity and acceleration. Open the experiment file called **Dueling Fan Units (L04A3-1)** to display the velocity and acceleration axes shown below.

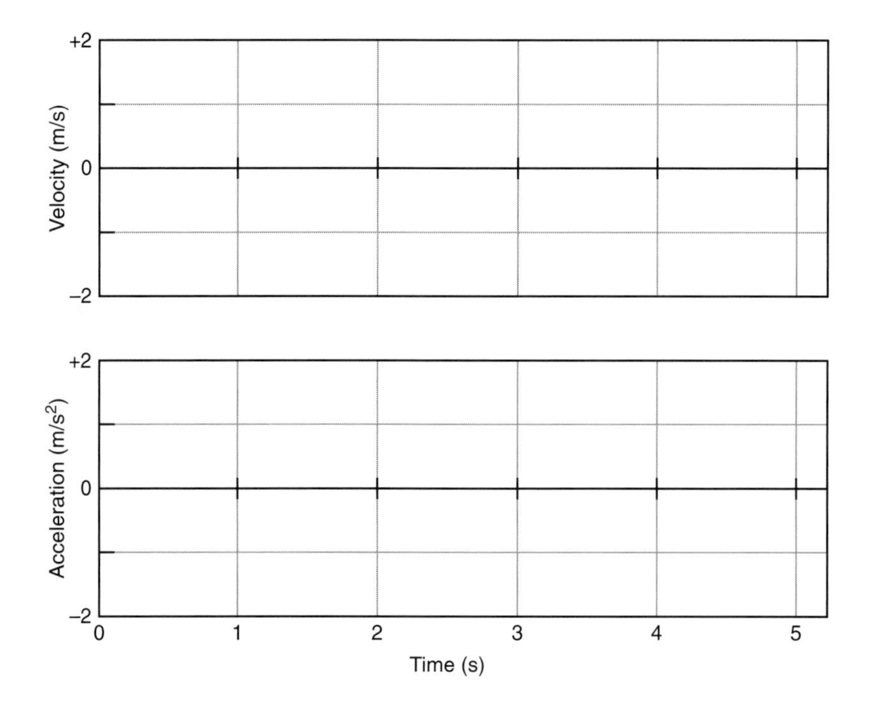

5. Now turn on both fan units and **begin graphing.** Give the cart a quick push toward the right and release it. Move the data so that the graphs are **persistently displayed on the screen** for comparison.

Question 3-2: How do the portions of these graphs after the cart was released compare to those for the motions in Activity 2-2 where the fan units both pushed in the same direction?

6. Turn on both fan units and give the cart a bigger push to the right. Sketch both sets of graphs on the previous axes with different lines, or **print** and affix over the axes.

Question 3-3: Compare the graphs with the larger push to those with the smaller push. Are there any differences in the velocity graphs? In the acceleration graphs?

7. Repeat, this time with a push to the left. Sketch your graphs on the axes with dotted lines, or **print** them.

Question 3-4: Compare the graphs to those with the push toward the right. Are there any differences in the velocity graphs? In the acceleration graphs?

Question 3-5: What combined force on the cart will cause it to move with a constant velocity? Explain based on your observations.

> **Comment:** In this activity you have looked at a situation where the combined force acting on the cart is zero. As you have seen, the velocity of the cart does not change. The cart either moves with a constant velocity or remains at rest. The law describing the motion when the combined force acting on an object is zero is known as *Newton's first law.* It is a special case of *Newton's second law* when the combined force is zero.

If you have more time, do the following Extension.

Extension 3-2: Once a Pull, Always a Pull?

You have seen in Activity 3-1 that to make the cart move with a constant velocity you needed to apply a force to get it moving, but no applied force (or a very small force to balance the frictional force) was needed to keep it moving at a constant velocity.

Prediction E3-4: Suppose that you remove the fan units and mount the force probe on the cart as in the diagram below.

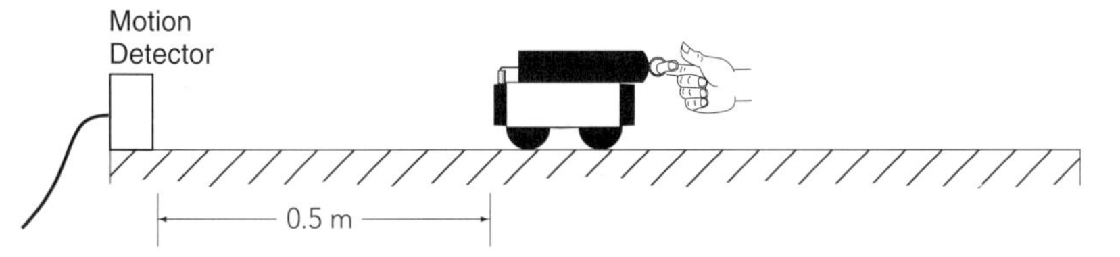

Next you give the cart a quick pull to start it moving away from the motion detector (by pulling on the force probe hook) and release it. Sketch with dashed lines on the axes that follow the velocity–time, acceleration–time and force–time graphs for the motion of the cart. Indicate the moment when the cart was released with an arrow.

Prediction E3-5: Describe the velocity, acceleration, and force in words in the space below.

Test your predictions. The cart's friction should be as small as possible. Be sure that the force probe cord won't interfere with the cart's motion and won't be seen by the motion detector. **Calibrate** the force probe with a 2.0 N pull from a spring scale or **load the calibration,** if this hasn't already been done in Investigation 2. As always, **zero** the force probe just before graphing. You may use the experiment file called **Once a Pull (L04E3-2)** to display the axes shown below.

Begin graphing. After the motion detector starts clicking, give the cart a short pull on the hook of the force probe in the direction away from the motion detector, and then let the cart go.

Sketch your graphs on the axes with solid lines, or **print** and affix them over the axes. Indicate with an arrow the time when the pull stopped.

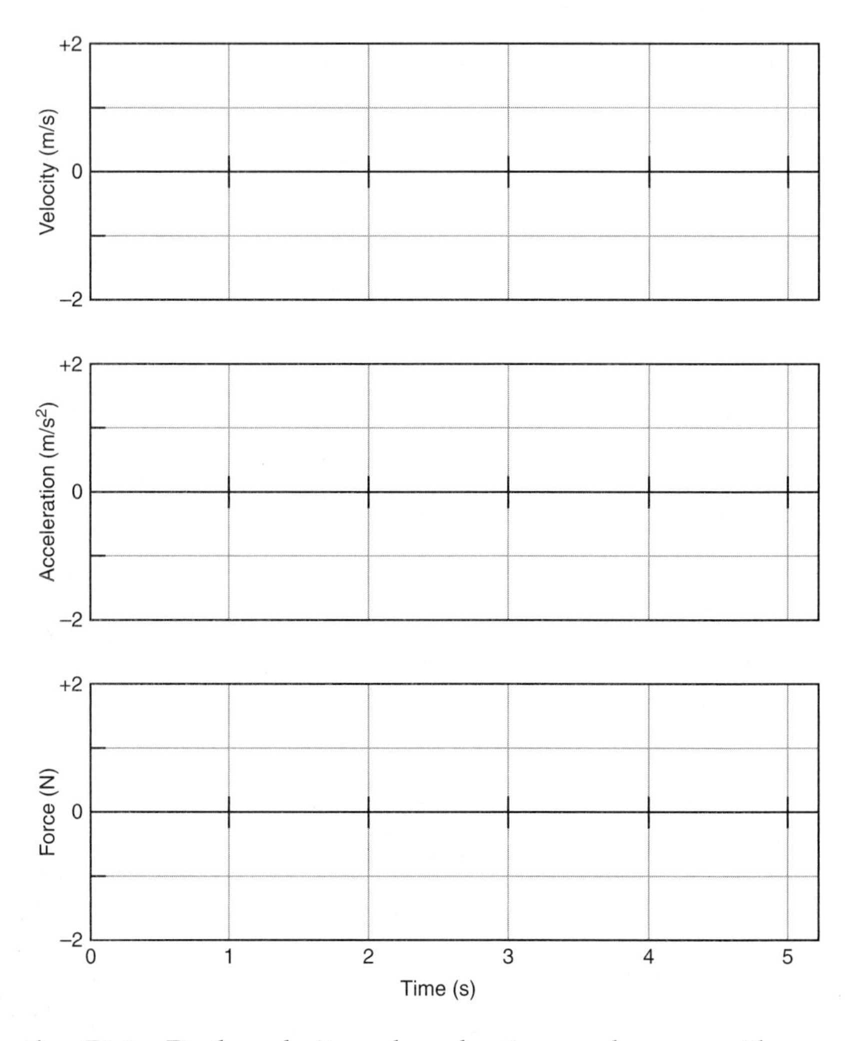

Question E3-6: Do the velocity and acceleration graphs agree with your predictions? If not, how do they differ? What happened to the force of the pull after you released the cart? What happened to the acceleration after you released the cart? Explain.

Question E3-7: Do your results agree with what you observed in Activity 3-1 about the applied force needed to keep an object moving at a constant velocity? Explain.

Name________________________ Date__________ Partners________________________

Homework for Lab 4: Combining Forces

Questions 1–5 refer to a toy car that can move in either direction along a horizontal line (the + position axis).

0 +

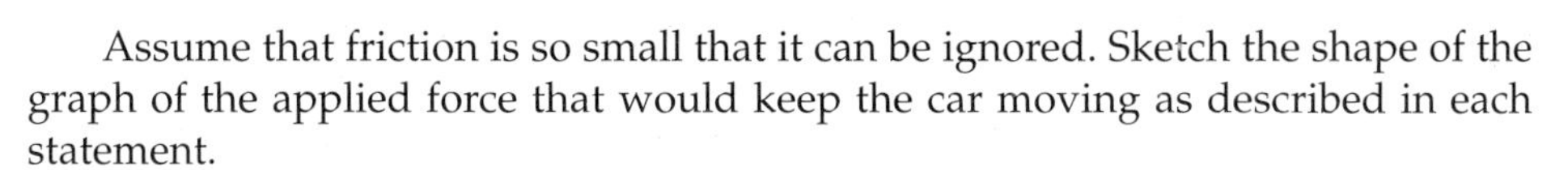

Assume that friction is so small that it can be ignored. Sketch the shape of the graph of the applied force that would keep the car moving as described in each statement.

1. The toy car moves away from the origin with a constant velocity.

2. The toy car moves toward the origin with a constant velocity.

3. The toy car moves away from the origin with a steadily decreasing velocity (a constant acceleration).

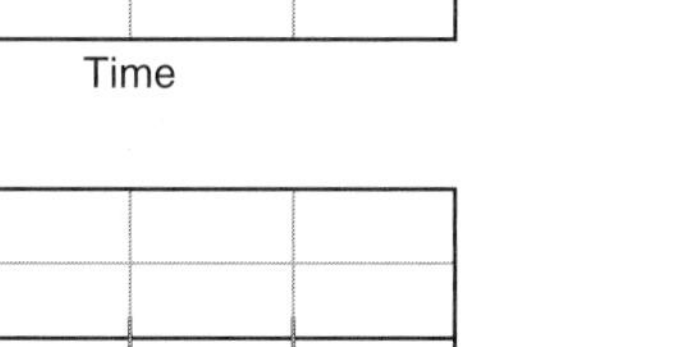

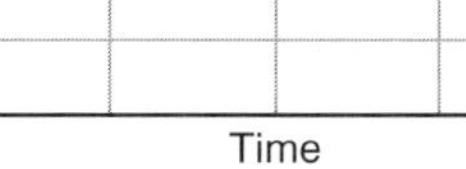

4. The toy car moves away from the origin, speeds up, and then slows down.

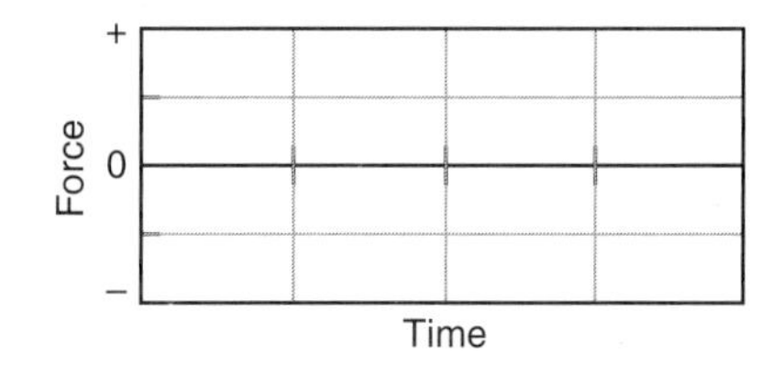

5. The toy car moves toward the origin with a steadily increasing velocity (a constant acceleration).

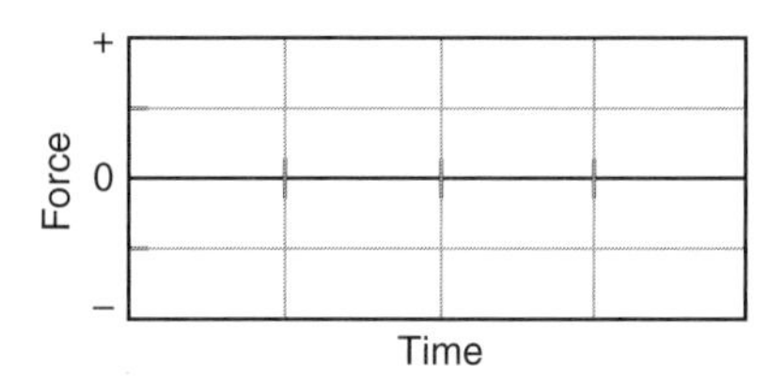

6. The toy car is given a push away from the origin and released. It continues to move with a constant velocity. Sketch the force after the car is released.

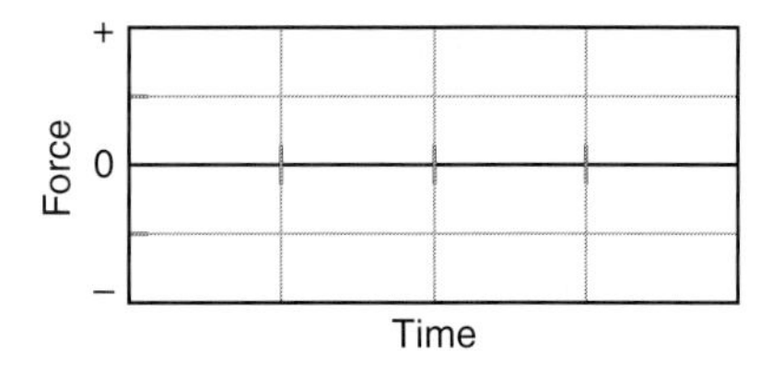

7. A cart is moving toward the right and speeding up, as shown in the diagram below. Draw arrows above the cart representing the magnitudes and directions of the net (combined) forces you think are needed on the cart at $t = 0$ s, $t = 1$ s, etc., to maintain its motion with a steadily increasing velocity. Assume that the cart is already moving at t_1.

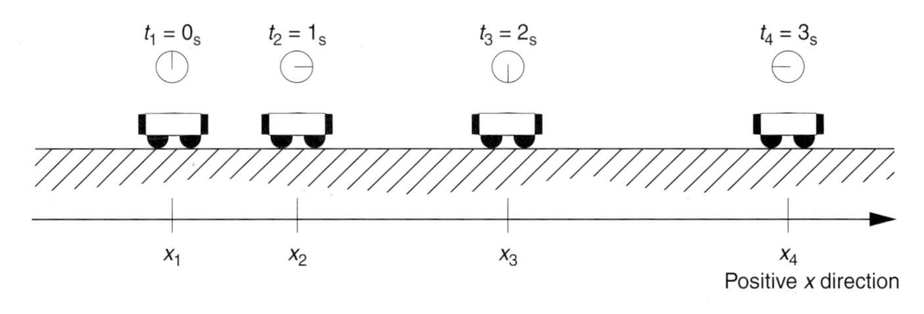

Explain the reasons for your answers.

8. If the positive direction is toward the right, what is the sign of the force at $t = 2$ s in Question 7? Explain.

9. A cart is moving toward the right and slowing down, as shown in the diagram below. Draw arrows above the cart representing the magnitudes and directions of the net (combined) forces you think are needed on the cart at $t = 0$ s, $t = 1$ s, etc., to maintain its motion with a steadily decreasing velocity. Assume that the cart is moving at t_1 and t_4.

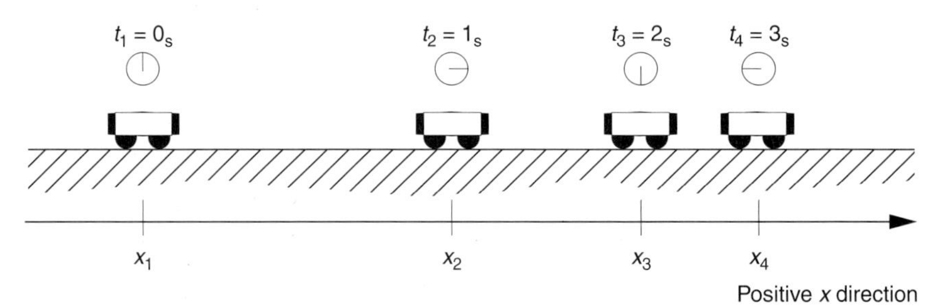

Explain the reasons for your answers.

10. If the positive direction is toward the right, what is the sign of the force at $t = 2$ s in Question 9? Explain.

11. A toy car can move in either direction along a horizontal line (the + position axis).

Assume that friction is so small that it can be ignored. A force toward the right of constant magnitude is applied to the car.

Sketch on the axes below using a solid line the shape of the acceleration–time graph of the car.

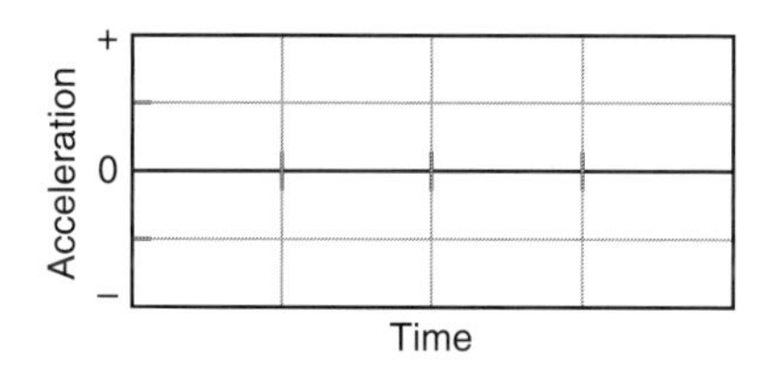

Explain the shape of your graph in terms of the applied force.

In Questions 12–15, assume that friction is so small that it can be ignored.

12. The spring scale in the diagram below reads 10.5 N.

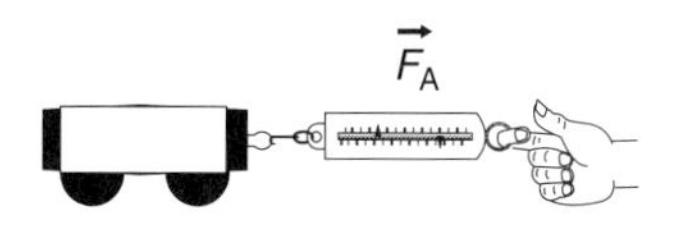

The cart moves toward the right with an acceleration toward the right of 3.25 m/s^2. Now two forces are applied to the cart with two different spring scales as shown below. The spring scale F_A still reads 10.5 N.

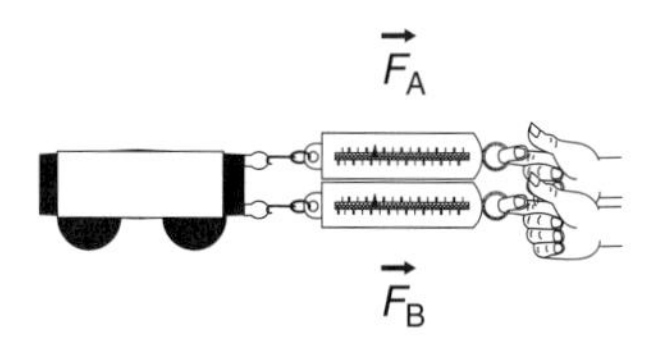

The cart now moves toward the right with an acceleration toward the right of 5.50 m/s^2. What does spring scale F_B read? Show your calculations, and explain.

13. Now two forces are applied to the cart with two different spring scales as shown below. The spring scale F_A still reads 10.5 N.

The cart now moves toward the right with an acceleration toward the right of 2.50 m/s^2. What does spring scale F_B read? Show your calculations, and explain.

14. Again two forces are applied to the cart with two different spring scales as shown below. The spring scale F_A still reads 10.5 N.

The cart moves with a constant velocity toward the right. What does spring scale F_B read? Show your calculations and explain.

15. Again two forces are applied to the cart with two different spring scales as shown below. The spring scale F_A still reads 10.5 N.

The cart moves toward the left with an acceleration toward the left of 2.50 m/s^2. What does spring scale F_B read? Show your calculations and explain.

Name______________________________ Date__________________

PRE-LAB PREPARATION SHEET FOR LAB 5: FORCE, MASS, AND ACCELERATION

(Due at the beginning of Lab 5)

Directions:
Read over Lab 5 and then answer the following questions about the procedures.

1. How will you find masses equal to numbers of "cart masses" in Activity 1-1?

2. Is the relationship between combined (net) applied force and acceleration for an object with constant mass a *direct, linear,* or *proportional* one?

3. Sketch a graph on the axes on the right of y vs. x where the mathematical relationship is inversely proportional.

y

x

4. How will you examine the relationship between acceleration and mass when you change the mass of an object while applying the same constant force?

5. If you apply a force of 1.0 N to a cart of mass 1.0 kg with negligible friction, what do you expect the acceleration to be?

Name______________________ Date____________ Partners______________________

Lab 5: Force, Mass, and Acceleration

". . . equal forces shall effect an equal change in equal bodies . . ."

—I. Newton

OBJECTIVES

- To develop a definition of mass in terms of an object's acceleration under the influence of a force.
- To find a mathematical relationship between the acceleration of an object and its mass when a constant force is applied—*Newton's second law.*
- To examine the mathematical relationship between force, mass, and acceleration—*Newton's second law*—in terms of the SI units (N for force, kg for mass, and m/s^2 for acceleration).
- To develop consistent statements of *Newton's first* and *second laws of motion* for one-dimensional motion (along a straight line) for any number of one-dimensional forces acting on an object.

OVERVIEW

In this lab you will continue to develop the first two of Newton's famous laws of motion. You will do this by combining careful definitions of force and mass with observations of the mathematical relationships among these quantities and acceleration.

You have seen that the acceleration of an object is directly proportional to the *combined* or net force acting on the object. If the combined force is not zero, then the object will accelerate. If the combined force is constant, then the acceleration is also constant. These observations can be summarized by *Newton's second law of motion.*

In Lab 4, you have also seen that for an object to move at a constant velocity (zero acceleration) when friction is negligible, the combined or net force on the object is zero. (You will see later that friction can be treated as a force and included in the calculation of net force.) The law that describes constant velocity motion of an object is *Newton's first law of motion.* Newton's first and second laws of motion are very powerful! They allow you to relate the net force on an object to its subsequent motion, and to make mathematical predictions of the object's motion.

What if a force were applied to an object having a larger mass? A smaller mass? How would this affect the acceleration of the object? In Investigation 1 of this lab you will study how the amount of "stuff" (mass) experiencing a force affects the magnitude of its acceleration.

In Investigation 2 you will study more carefully the definitions of the units in which we express force, mass, and acceleration.

INVESTIGATION 1: FORCE, MASS AND ACCELERATION

In previous activities you have applied forces to a cart having the same mass in each case and examined its motion. But when you apply a force to an object, you know that the object's mass has a significant effect on its acceleration. For example, compare the different accelerations that would result if you pushed a 1000-kg (metric ton) automobile and a 1-kg cart, with the same force!

In this investigation you will explore the mathematical relationship between acceleration and mass when you apply the same constant force to carts of different mass.

You will need

- computer-based laboratory system
- *RealTime Physics Mechanics* experiment configuration files
- force probe
- motion detector
- low-friction cart of mass about 0.5 kg
- variety of masses to increase the mass of the cart, totaling 2–3 times the mass of the cart
- spring scale with a maximum reading of 5 N
- equal arm (two pan) balance
- smooth ramp or other level surface 2–3 m long
- low-friction pulley and string
- variety of hanging masses (10–50 g)

Activity 1-1: Acceleration and Mass

You can easily change the mass of the cart by attaching masses to it, and you can apply the same force each time by using a string attached to appropriate hanging masses. By measuring the acceleration of different mass carts, you can find a mathematical relationship between the acceleration of the cart and its mass, when the force applied by the string is kept constant.

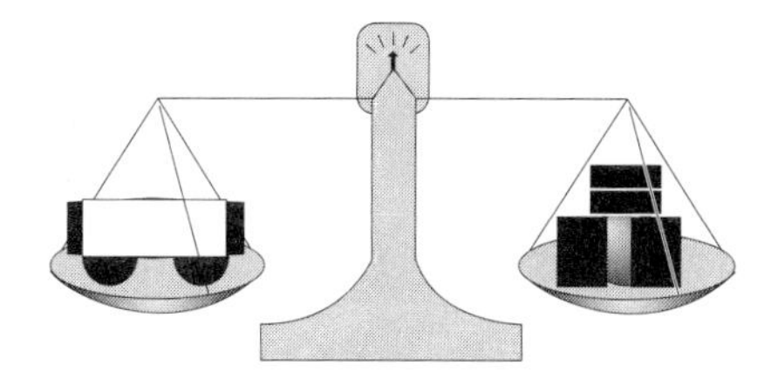

1. Set up the ramp, pulley, cart, string, motion detector, and force probe as shown in the figure that follows. *Be sure that the ramp is level.*

   The force probe should be fastened securely to the cart. *(Be sure that the force probe does not extend beyond the end of the cart.* The cable must not interfere with the motion of the cart and must not be seen by the motion detector.)

2. We will define a mass scale in which the unit is the mass of the cart *(including the force probe),* called one *cart mass.* An equal arm balance can be used to assemble a combination of masses equal to *one cart mass.* If this combination of masses is divided in half, each half is 0.5 cart mass.

   Use the balance in this way to assemble masses that you can add to the cart to make the cart's mass equal to 1.5, 2.0, 2.5, and 3.0 cart masses. Label these masses.

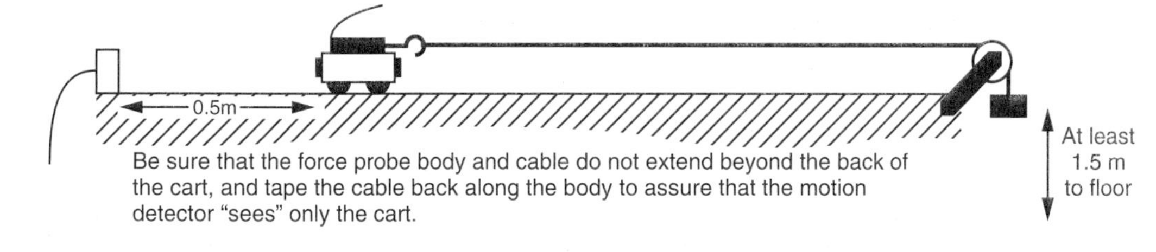

3. Begin by adding enough masses to make the cart's mass 2.0 cart masses.

4. Be sure that the cart's friction is minimum. (If the cart has a friction pad, it should be raised so that it doesn't contact the ramp.)

5. Find a hanging mass that will accelerate the 2.0 cart mass cart across the track from left to right in about 2–3 s as it is falling.

   Record the value of this mass:______kg

6. **Calibrate** the force probe with a 2.00-N pull or **load the calibration.** (If you are using a Hall effect force probe, you may need to adjust the spacing and **check the sensitivity.**)

7. Open the experiment file called **Acceleration & Mass (L05A1-1)** to display the axes that follow.

8. As always, **zero** the force probe before each graph with nothing pulling on it. **Begin graphing.** Release the cart from rest when you hear the clicks of the motion detector.

   Move your data so that the graph is **persistently displayed on the screen** for later comparison. (You will probably want to wait to print until you have completed all graphing in Activity 1-2 and Extension 1-3.)

   Sketch your graphs on the axes that follow, or **print** them and affix them over the axes.

9. Use arrows to mark the time interval during which the acceleration is nearly constant on your graph. Use the **analysis and statistics features** of the software to measure the average force experienced by the cart and average acceleration *during the same time interval.* Record your measured values for average force and average acceleration in the third row of Table 1-1.

Table 1-1

Mass of cart (cart masses)	Average applied force (N)	Average acceleration (m/s^2)
1.0		
1.5		
2.0		
2.5		
3.0		

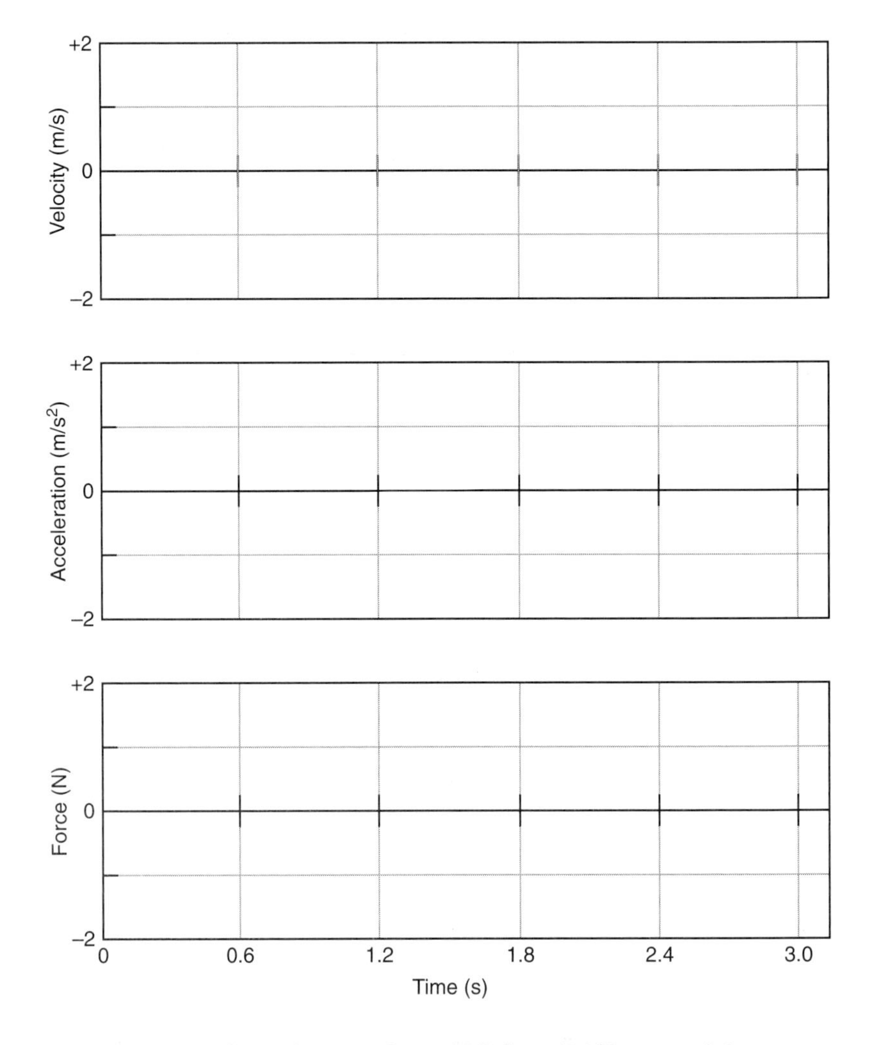

Activity 1-2: Accelerating a Cart With a Different Mass

Prediction 1-1: Suppose that you remove mass from the cart until it is 1.0 cart mass, and accelerate it *with the same applied force*. Compare the acceleration to that of the larger mass cart.

1. Test your prediction. Remove the masses you added in Activity 1-1 that doubled the mass of the cart to 2.0 cart masses.

Comment: You want to accelerate the cart with the *same applied force.* As you may have noticed, the force applied to the force probe by the string decreases once the cart is released. (You will explore why this is so in a later lab.) This decrease depends on the size of the acceleration. Therefore, in order to keep the applied force constant, you may need to change the hanging mass slightly.

2. **Zero** the force probe with no force on it. Adjust the hanging mass until the force probe reading *while the cart is accelerating* is the same as the force you recorded in the third row of Table 1-1.

   When you have found the correct hanging mass, graph the motion of the cart. (Don't forget to **zero** the force probe first.) Measure the average force and average acceleration of the cart *during the time interval when the force and acceleration are nearly constant* and record these values in the first row of Table 1-1.

Question 1-1: Did the acceleration agree with your prediction? Explain.

3. Now make the mass of the cart 1.5 cart masses, and accelerate it again *with the same size force.* (Don't forget to adjust the hanging mass, if necessary.) Measure the average force and acceleration of the cart, and record these values in the table.
4. Repeat for masses of 2.5 and 3.0 cart masses.

If you have more time, do the following Extension now, and take additional data for larger masses of the cart.

Extension 1-3: More Data for Larger Masses

Find the average acceleration for the cart with the same average applied force but for masses larger than 3.0 cart masses. Include these data in your graph in Activity 1-4.

Activity 1-4: Relationship Between Acceleration and Mass

1. Plot a graph of average acceleration vs. cart mass (with constant applied force). You can do this by opening the experiment file called **Avg. Accel. vs. Mass (L05A1-4).** Enter the acceleration and mass data into the table on the screen. You may wish to **adjust the axes** to better display the data.

Question 1-2: Does the acceleration of the cart increase, decrease, or remain the same as the mass of the cart is increased?

Comment: We are interested in the nature of the mathematical relationship between average acceleration and mass of the cart, with the applied force kept constant. As always, this can be determined from the graph by drawing a smooth curve which fits the plotted data points.

Some definitions of possible mathematical relationships when y decreases as x increases are shown in the sketches below. In these examples, y might be the average acceleration, and x the mass of the cart.

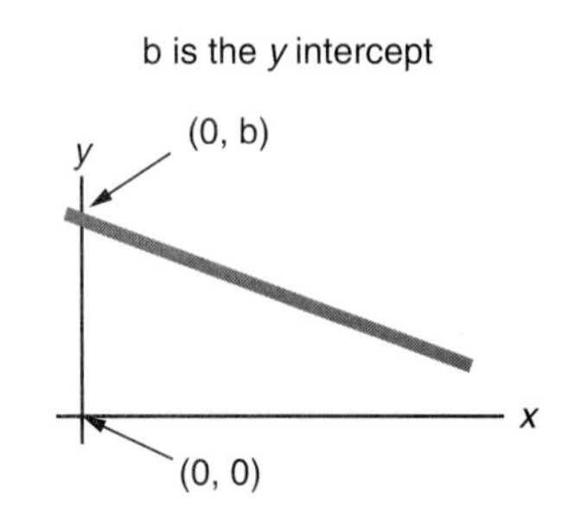

y is a *linear* function, which decreases linearly as x increases according to the mathematical relationship $y = mx + b$, where the slope m is a negative constant and $b =$ const.

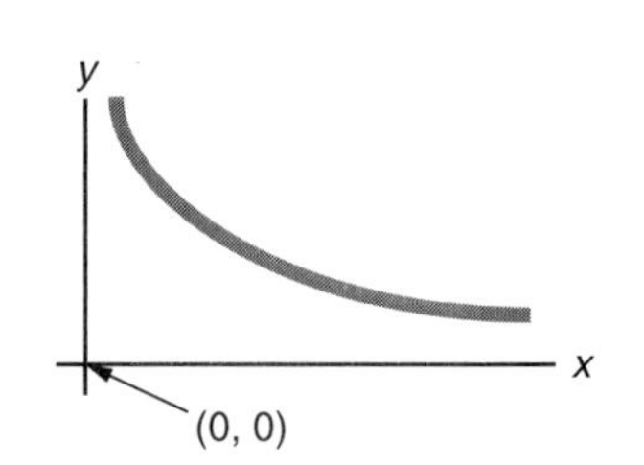

y is a function of x, which decreases as x increases. The mathematical relationship might be $y = b + cx^{-n}$, where n is an integer. When $b = 0$ and $n = 1$, the relationship becomes $y = c/x$, and y is said to be *inversely proportional* to x.

Note that these *are not all the same.* y can decrease as x increases, and the relationship doesn't have to be *linear* or *inversely proportional. Inverse proportionality* refers *only* to the special relationship where $y = c/x$, where c is a constant. The motion software allows you to determine the relationship by trying various curves to see which best fits the plotted data.

2. Use the **fit routine** in the software to fit the data on your graph of average acceleration vs. mass of the cart. Select various possible relationships and test them.

3. When you have found the best fit, **print** the graph along with the fit equation and affix it below.

Question 1-3: What appears to be the mathematical relationship between acceleration and mass of the cart, when the applied force is kept constant?

Question 1-4: In the previous lab, you found that the acceleration of the cart was proportional to the *combined* applied force when the mass of the cart was not changed. State in words the general relationship between the applied force, the mass, and the acceleration of the cart that you have found in these two labs. If the combined (net) force is $\Sigma\vec{F}$, the mass is m, and the acceleration is $\vec{a}$, write a mathematical relationship that relates these three physical quantities.

If you have more time, do the following Extension.

Extension 1-5: Acceleration vs. Mass for a Different Applied Force

Prediction E1-2: Suppose that you applied a larger force to the cart by using a hanging mass twice as large as before. How would the relationship between average acceleration and mass compare to that in Activity 1-4? In what ways would it be similar and in what ways would it be different?

Test your prediction. Repeat Activities 1-1 through 1-4, this time using a hanging mass twice as large to accelerate the cart.

Plot a graph of average acceleration vs. mass, as in Activity 1-4 and determine the relationship between acceleration and mass.

Question E1-5: Compare the relationship between acceleration and mass with this larger constant applied force to the one for the smaller applied force you used before. Is this what you expected?

INVESTIGATION 2: FORCE AND MASS UNITS

So far you have been measuring force in *standard* units based on the pull exerted by a spring scale calibrated in newtons. Where does this unit come from? By contrast, we have our own private units for measuring mass—cart masses. If one group were using a large wooden cart in their force and motion experiments and another group were using a small aluminum cart with smaller mass, they would have different values for mass and would observe different accelerations for "one cart mass pulled by one newton." It's time to discuss standard units for force and mass.

It would be nice to be able to do a mechanics experiment in one part of the world and have scientists in another part of the world be able to replicate it or at least understand what actually happened. This requires that people agree on standard units. In 1960 an international commission met to develop a common set of units for fundamental quantities such as length, time, mass, force, electric current, and pressure. This commission agreed that the most fundamental units in the study of mechanics are those of length, time, and mass. All other units, including those of force, work, energy, torque, and rotational velocity, that you encounter in your study of mechanics can be expressed as a combination of these basic quantities. The fundamental *International System* or *SI* units along with the standard unit for force are shown in the boxes below.

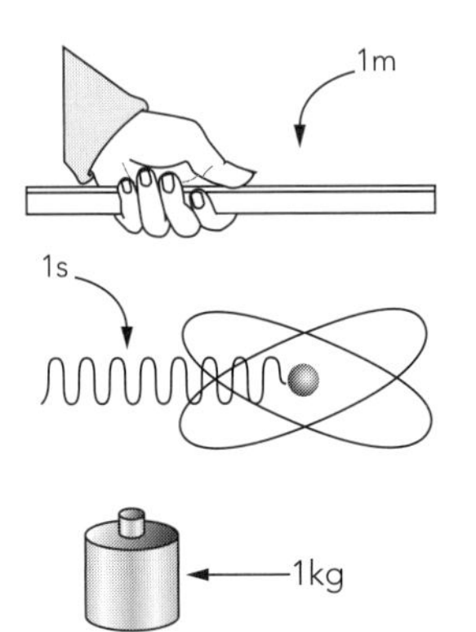

FUNDAMENTAL UNITS FOR MECHANICS

Length: A **meter (m)** is the distance traveled by light in a vacuum during a time of 1/299,792,458 s.

Time: A **second (s)** is defined as the time required for a cesium-133 atom to undergo 9,192,631,770 vibrations.

Mass: A **kilogram (kg)** is defined as the mass of a platinum–iridium alloy cylinder kept at the International Bureau of Weights and Measures in Sévres, France. It is kept in a special chamber to prevent corrosion.

$1 \text{ N} = 1 \text{ kg}\cdot\text{m/s}^2$

THE FORCE UNIT EXPRESSED IN TERMS OF LENGTH, MASS, AND TIME

Force: A **newton (N)** is defined as that force which causes a 1-kg mass to accelerate at 1 m/s^2.

We want to be able to measure masses in kilograms and forces in newtons in our own laboratory. The following activities are designed to give you a feel for standard mass and force units and how they are determined in the laboratory. You will need the following equipment:

- computer-based laboratory system
- *RealTime Physics Mechanics* experiment configuration files
- force probe
- motion detector
- balance or electronic scale to measure masses in kilograms
- spring scale with a maximum reading of 5 N
- low-friction cart
- smooth ramp or other level surface 2–3 m long
- low-friction pulley and string
- assortment of hanging masses (10–50 g)

Our approach in the following activities is to use a standard force scale to calibrate the force probe in newtons. Using your data from the previous investigation, you will see how you can establish a mass unit in terms of your force and acceleration measurements. Then you will use a standard mass scale to get enough stuff loaded on a cart to equal one kilogram of mass. Finally, you can pull the cart with a force of about one newton and see if it accelerates at something close to one meter per second squared.

WARNING! There will probably be a noticeable amount of uncertainty associated with your measurements.

Suppose you want to find the mass of an object in kilograms. You need to compare it to the 1-kg platinum–iridium alloy cylinder at the International Bureau of Weights and Measures in France. It would be nice to have a standard kilogram in your laboratory. You could go to France, but it is unlikely that they would let you take the standard home with you!

Suppose, however, that you go to France and accelerate the *standard* mass with a constant force and measure the force and also the resulting acceleration as accurately as possible. Next you would need to make a cylinder that seemed just like the standard one and add or subtract stuff from it until it undergoes *exactly the same acceleration with the same constant force.* Then within the limits of experimental uncertainty this new cylinder standard and the bureau standard would have the same mass. If the comparison could be made to three significant figures, then the mass of your new standard would be $m_{std} = 1.00$ kg.

Suppose you head home with your standard mass. You wish to determine the mass of another object. You could apply the same constant force F on the standard and on the other object, and measure both accelerations. Then, according to *Newton's second law,* $F = ma$,

$$m_{std} = 1.00 \text{ kg} = \frac{F}{a} \qquad m_{other} = \frac{F}{a_{other}}$$

Since the constant force, F, applied to both masses was the same,

$$m_{other} = 1.00 \text{ kg} \frac{a}{a_{other}}$$

In fact, you already did something similar in the last investigation.

Activity 2-1: Calculating One "Cart Mass" in Standard Units

1. In Investigation 1 of this lab, you measured the force applied to a cart and the acceleration of the cart with mass equal to 1.0, 1.5, 2.0, 2.5, and 3.0 cart masses. Turn back to Table 1-1 from that experiment and copy the values of average force and average acceleration into the second and third columns of Table 2-1.

Table 2-1

Mass of cart (cart masses)	Average applied force (N)	Average acceleration (m/s^2)	Ratio of F/a (calculated mass)	Mass of cart measured with balance (kg)
1.0				
1.5				
2.0				
2.5				
3.0				

In the discussion above, the mass in standard units was calculated using Newton's second law by taking the ratio of the combined (net) force on the object in newtons to the acceleration of the object measured in meters per second squared.

2. For each row in Table 2-1, calculate the ratio of the force to acceleration and record it in the fourth column.

Question 2-1: According to the discussion, the values you just calculated should be the masses of the objects in kilograms. Do your numbers seem to make sense? What do you get for the value of 1.00 cart mass in kilograms? What do you get for the value of 2.00 cart masses in kilograms?

Comment: Physicists call the quantity you have just calculated—the ratio of combined (net) force on an object to its acceleration—the *inertial mass* of the object.

You could continue to determine and compare masses by accelerating them and taking force to acceleration ratios, but this process is pretty tedious. A simpler approach is to use an electronic scale or a mechanical balance that has already been calibrated in kilograms by somebody who is intelligent and knowledgeable using a standard mass! (The details of why such devices can give us correct masses in kilograms will not be easy to understand fully until after gravitational forces are studied in Lab 6.)

3. Compare your inertial mass calculations for 1.0, 1.5, 2.0, 2.5, and 3.0 cart masses with the values you get by placing your cart on an electronic scale or mechanical balance. Record these values in the last column of Table 2-1.

Question 2-2: Are your inertial masses reasonably consistent with your masses measured with the scale or balance?

Comment: In your experiments, you have seen that the physical quantities force, mass and acceleration are related through *Newton's second law.* In the activity you have just done, you have used this relationship to *define* inertial mass in terms of *standard* units of force, length, and time. This is a good logical definition of inertial mass.

Historically, however, the units of mass, length, and time were defined first as *standards* and the unit of force was defined as a *derived* unit in terms of these standard units. Thus, *a newton of force is defined as the force needed to accelerate 1.00 kg at 1.00 m/s*^{2}. In the next activity you will examine this definition.

Activity 2-2: Does a Force of 1.0 N Applied to a 1.0-kg Mass Really Cause an Acceleration of 1.0 m/s^2?

You have used mass and force measuring devices that have been provided for you. You can now see if everything makes sense by accelerating one kilogram of mass with a force of about one newton and seeing if an acceleration of about one meter per second squared results.

1. Set up the ramp, pulley, weighted cart, string, motion detector, and force probe as in Activity 1-1.

   Tape masses to the cart along with the force probe so that the total mass of the cart is 1.0 kg.

   Be sure that the cable from the force probe doesn't interfere with the motion of the cart and is out of the way of the motion detector.

2. Open the experiment file called **Acceleration with 1.0 N Force (L05A2-2)** to set up axes to graph velocity, acceleration, and force.

3. **Calibrate** the force probe with a force of 2.0 N using the spring scale or **load the calibration,** if it hasn't already been calibrated. (If you are using a Hall effect force probe, be sure to **check the sensitivity** of the force probe before you calibrate.)

4. Remember to **zero** the force probe with nothing pulling on it before each run. Measure the acceleration that results from a 1.0-N force applied to the force probe. Try different hanging masses until you get an applied force of close to 1.0 N *while the cart is accelerating.*

Comment: *Be careful!* Remember that when the cart is being held at rest, the same hanging mass will exert more applied force on the cart than when it is accelerating.

5. Once you get a good run, use the **analysis and statistics features** of the software to measure the average values of force and acceleration, and record these values in the table below. Also record the hanging mass.

Mass of cart (kg)	
Average applied force (N)	
Average acceleration (m/s^2)	
Hanging mass (kg)	

Question 2-3: How close is your result to the expected value of acceleration—1.0 m/s^2? Discuss sources of uncertainty in your measurements of acceleration and force.

Question 2-4: A force of 5.4 N is applied to an object, and the object is observed to accelerate with an acceleration of 3.0 m/s^2. If friction is so small that it can be ignored, what is the mass of the object in kilograms? Show your calculation.

Question 2-5: An object of mass 39 kg is observed to accelerate with an acceleration of 2.0 m/s^2. If friction is so small that it can be ignored, what is the force applied to the object in newtons? Show your calculation.

Comment: The main purpose of Labs 3, 4, and 5 has been to explore the relationship between the forces on an object, its mass, and its acceleration. You have been developing *Newton's first and second laws of motion* for one-dimensional situations in which all forces lie in a positive or negative direction along the same line.

Activity 2-3: Newton's Laws in Your Own Words

Question 2-6: Express *Newton's first law* (the one about constant velocity) in terms of the *combined (net) force* applied to an object in your own words clearly and precisely.

Question 2-7: Express *Newton's first law* in equations in terms of the acceleration vector, the *combined (net)* force vector applied to an object, and the object's velocity.

$$\text{If } \vec{F}^{\text{net}} = \Sigma\vec{F} = \quad \text{then } \vec{a} = \qquad \text{and } \vec{v} =$$

Question 2-8: Express *Newton's second law* (the one relating force, mass, and acceleration) in your own clear and precise words in terms of the *combined (net) force* applied to an object.

Question 2-9: Express *Newton's second law* in equations in terms of the acceleration vector, the *combined (net) force* vector applied to an object, and its mass.

$$\text{If } \vec{F}^{\text{net}} = \Sigma\vec{F} \neq 0, \qquad \text{then } \vec{a} =$$

Comment: The use of the equal sign in the mathematical representation of Newton's second law does not signify that an acceleration is the same as or equivalent to a force divided by a mass, but instead it spells out a procedure for calculating the magnitude and direction of the acceleration of a mass while it is experiencing a net force. What we assume when we subscribe to *Newton's second law* is that a net force on a mass *causes* an acceleration of that mass.

Name________________________ Date____________ Partners________________________

Homework for Lab 5: Force, Mass and Acceleration

1. Given the table of data below for widgets and doodads, how would you determine whether or not the relationship between widgets and doodads is an inversely proportional one? Sketch on the axes on the right of the table what the graph would look like if widgets are inversely proportional to doodads, and write the form of the equation that relates widgets to doodads in this case.

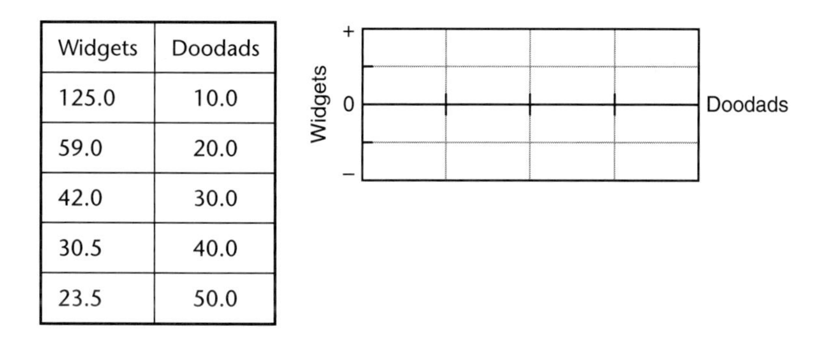

Widgets	Doodads
125.0	10.0
59.0	20.0
42.0	30.0
30.5	40.0
23.5	50.0

Questions 2-3 refer to a toy car that can move in either direction along a horizontal line (the + position axis).

Assume that friction is so small that it can be ignored. A force toward the right of constant magnitude is applied to the car.

2. Sketch on the axes below using a solid line the shape of the car's acceleration–time graph.

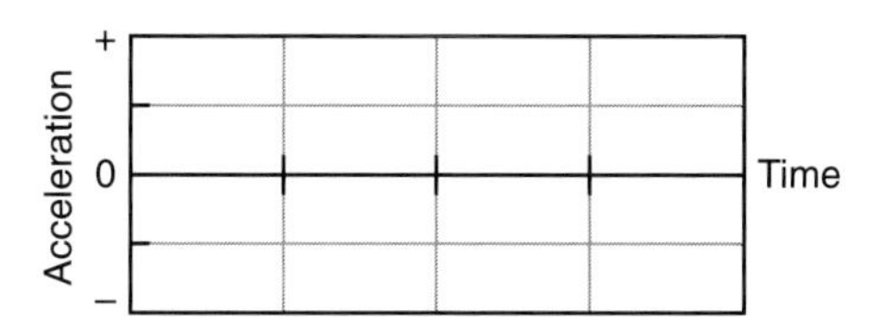

3. Suppose that the mass of the car were twice as large. The same constant force is applied to the car. Sketch on the axes above using a dashed line the car's

acceleration–time graph. Explain any differences in this graph compared to the car's acceleration–time graph with the original mass.

4. When a force is applied to an object with mass equal to the standard kilogram, the acceleration of the mass is $3.25\ \mathrm{m/s^2}$. (Assume that friction is so small that it can be ignored.) When the same magnitude force is applied to another object, the acceleration is $2.75\ \mathrm{m/s^2}$. What is the mass of this object? What would the second object's acceleration be if a force twice as large were applied to it? Show your calculations.

5. Given an object with mass equal to the standard kilogram, how would you determine if a force applied to it has magnitude equal to one newton? (Assume that frictional forces are so small that they can be ignored.)

6. Why is it necessary to calibrate a force probe? Describe how this is done.

In Question 7, assume that friction is so small that it can be ignored.

7. The spring scale in the diagram below reads 10.5 N.

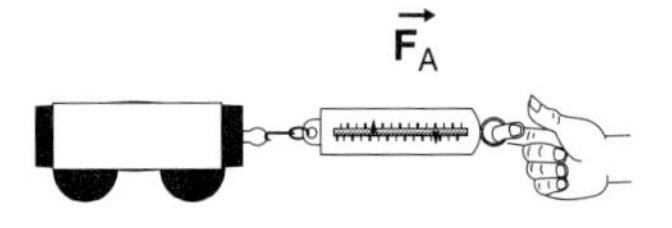

If the cart moves toward the right with an acceleration also toward the right of $3.25\ \mathrm{m/s^2}$, what is the mass of the cart? Show your calculations and explain.

In Questions 8–10, friction may *not* be ignored.

8. The force applied to the cart in Question 7 by spring scale F_A is still 10.5 N. The cart now moves toward the right with a constant velocity. What are the magnitude and direction of the frictional force? Show your calculations and explain your reasoning.

9. The force applied to the cart in Question 7 by spring scale F_A is still 10.5 N. The cart now moves toward the right with an acceleration also toward the right of 1.75 m/s^2. What are the magnitude and direction of the frictional force? Show your calculations and explain.

10. The force applied to the cart by spring scale F_A is 10.5 N.

The cart now moves toward the right with a constant velocity. The frictional force has the same magnitude as in Question 9. What does spring scale F_B read? Show your calculations and explain.

Name____________________________ Date________________

Pre-Lab Preparation Sheet for Lab 6: Gravitational Forces

(Due at the beginning of Lab 6)

Directions:
Read over Lab 6 and then answer the following questions about the procedures.

1. In Activity 1-1, why is the motion detector set up so that distance away from it is *negative*?

2. In Activity 1-1, why do you need to be careful about how you drop the ball?

3. How do you expect your velocity–time and acceleration–time graphs for Activities 1-1 and 1-4 will compare? Do you expect that they will be different or the same?

4. What is your Prediction 2-1? Will one ball have a larger acceleration, or will they both have the same acceleration?

5. What will you use to explore normal forces in Investigation 3?

Name________________________ Date____________ Partners________________________

Lab 6: Gravitational Forces

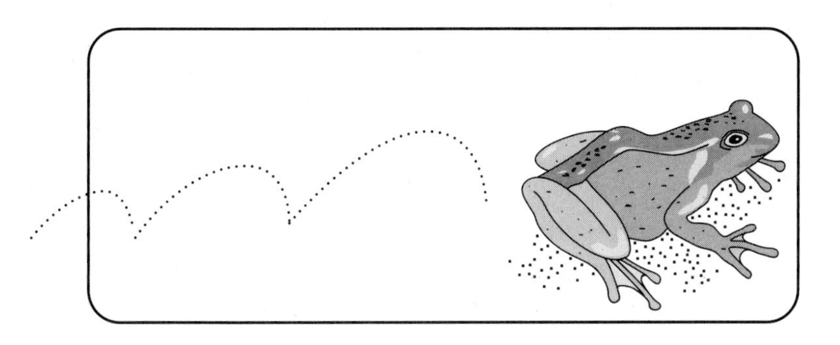

And thus Nature will be very conformable to herself and very simple, performing all the great Motions of the heavenly Bodies by the attraction of gravity . . .

—Isaac Newton

OBJECTIVES

- To explore the nature of motion along a vertical line near the Earth's surface.
- To extend the explanatory power of *Newton's laws* by inventing an *invisible* force (the gravitational force) that correctly accounts for the falling motion of objects observed near the Earth's surface.
- To examine the magnitude of the acceleration of a falling object under the influence of the gravitational force near the Earth's surface.
- To examine the motion of an object along an inclined ramp under the influence of the gravitational force.
- To define weight and discover its relationship to mass.
- To discover the origin of *normal* forces.

OVERVIEW

You started your study of *Newtonian dynamics* in Lab 3 by developing the concept of force. Initially, when asked to define forces, most people think of a *force* as an *obvious push or pull,* such as a punch to the jaw or the tug of a rubber band. By studying the acceleration that results from a force when little friction is present, we came up with a second definition of *force* as *that which causes acceleration.* These two alternative definitions of force do not always appear to be the same. Pushing on a wall doesn't seem to cause the wall to move. An object dropped close to the surface of the Earth accelerates and yet there is no visible push or pull on it.

The genius of Newton was to recognize that he could define *net force or combined force* as that which causes acceleration, and that if the obvious applied forces did not account for the degree of acceleration then there must be other "invisible" forces present. A prime example of an invisible force is the gravitational force—the attraction of the Earth for objects.

When an object falls close to the surface of the Earth, there is no obvious force being applied to it. Whatever is causing the object to move is invisible. Most people rather casually refer to the cause of falling motions as the action of "gravity." What is gravity? Can we describe gravity as just another force? Can we describe its effects mathematically? Can *Newton's laws* be interpreted in such a way that they can be used for the mathematical prediction of motions that are influenced by gravity?

In this lab you will first study vertical motion and the gravitational force. Then you will look at the motion of an object along an inclined ramp. You will also explore the relationship between mass and weight, and the meaning of mass.

Later you will examine the mechanism for *normal* force—a common type of force that often opposes the gravitational force to keep an object from moving.

INVESTIGATION 1: MOTION AND GRAVITY

Let's begin the study of the phenomenon of gravity by examining the motion of an object such as a ball when it is allowed to fall vertically near the surface of the Earth. This study is not easy, because the motion happens very quickly! You can first predict what kind of motion the ball undergoes by tossing a ball in the laboratory several times and seeing what you think is going on.

A falling motion is too fast to observe carefully by eye. You will need the aid of the motion detector and computer to examine the motion quantitatively. Fortunately, the motion detector can do measurements just fast enough to graph this motion. To carry out your measurements you will need

- computer-based laboratory system
- *RealTime Physics Mechanics* experiment configuration files
- motion detector
- tape or other mechanism to attach motion detector to the ceiling
- basketball or other uniformly round large ball

Activity 1-1: Motion of a Falling Ball

Prediction 1-1: Toss a ball straight up a couple of times and observe its motion *as it falls downward.* Describe in words how you think it might be moving. Some possibilities include falling at a constant velocity, falling with an increasing acceleration, falling with a decreasing acceleration, or falling with a constant acceleration. What do you think? Explain how you based your prediction on your observations of the ball's motion.

Prediction 1-2: Suppose that you drop the ball from a height of about 2 m above the floor, releasing it from rest. On the axes that follow, sketch your predictions for the velocity–time and acceleration–time graphs of the ball's motion. Assume that the positive y direction is *upward.*

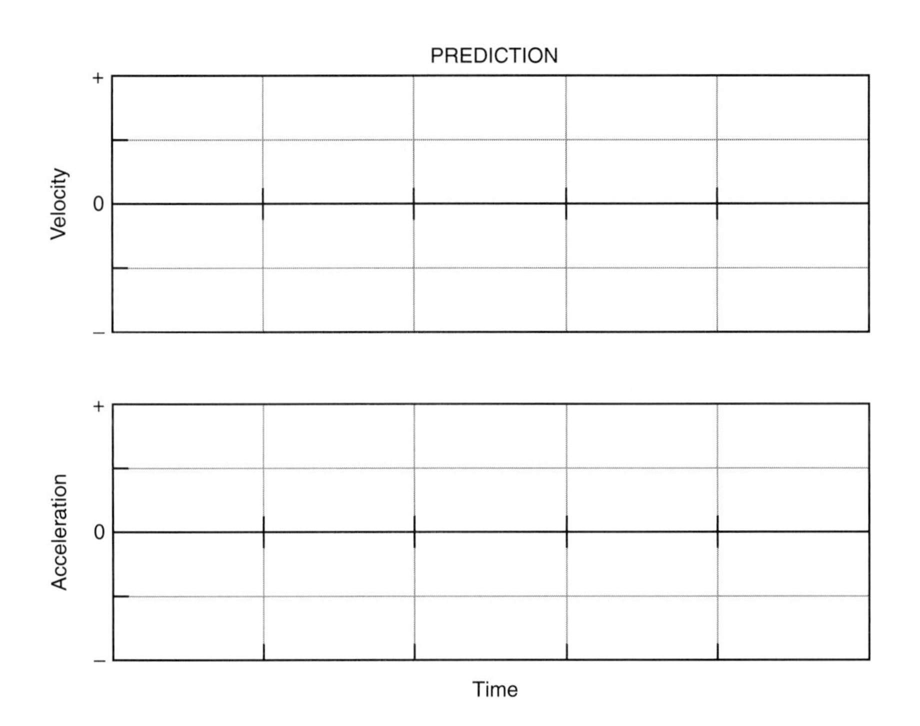

Test your predictions. You have previously used the motion detector to examine the motion of your body and of a cart. The motion of a falling ball takes place much faster. While you can use the motion detector to measure position, velocity, and acceleration, you will have to collect data at a faster rate than you have before.

The easiest way to examine the motion of the falling ball is to mount the motion detector as high up as you can and to use a large ball that is not too light (like a basketball rather than a beachball). It is essential to keep your hands and the rest of your body out of the way of the motion detector after the ball is released. This will be difficult and may take a number of tries. It also will take some care to identify which portions of your graphs correspond to the actual downward motion of the ball and which portions are irrelevant.

1. Tape the motion detector to a light fixture or something else as high above the floor as possible, with the detector looking straight downward, as shown on the right.

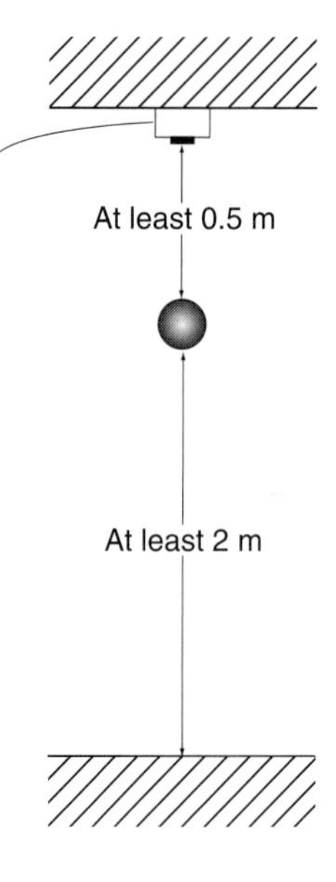

2. Set up the axes that follow by opening the experiment file called **Falling Ball (L06A1-1).** This will also set the data collection to a faster rate than has been used before (30 points/s).

   Because the motion detector is pointing downward—in the negative y direction—the software has been set up to call **distance away from the detector negative.**

3. Hold the ball at least 0.5 m directly below the motion detector and at least 2 m above the floor. *Remember that your hands and body must be completely out of the path of the falling ball, so the detector will see the ball—and not your hands or body—the whole way down.*

4. When everything is ready, **begin graphing** and release the ball as soon as you hear the clicks from the motion detector.

5. **Adjust the axes** if necessary to display the velocity and acceleration as clearly as possible.

   Move the data so that the graphs are **persistently displayed on the screen** for comparison in Activity 1-3.

   Sketch your graphs on the axes that follow, or **print** them and affix them over the axes.

6. Mark with arrows the beginning and end of the time interval during which the ball was falling freely.

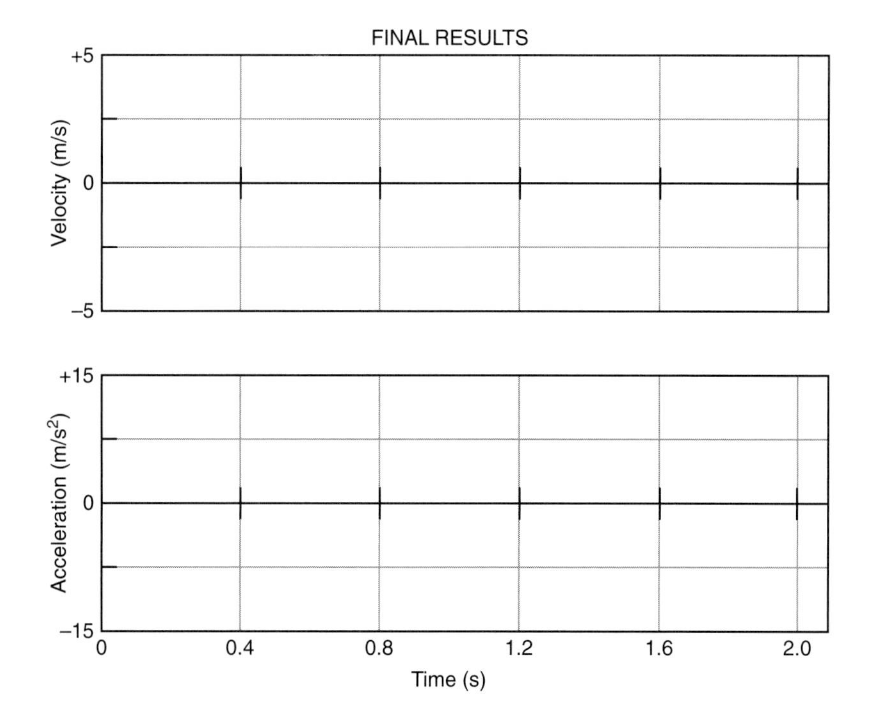

Question 1-1: What does the nature of the motion look like—constant velocity, constant acceleration, increasing acceleration, decreasing acceleration, or other? How do your observations compare with the predictions you made?

Question 1-2: What motion of a different object observed in a previous lab resulted in similar graphs to the ones for the falling ball? Describe what was moving and precisely how it was moving.

Question 1-3: Is the acceleration of the ball positive or negative as it falls down? Does this sign agree with the way that the velocity appears to be changing on the velocity–time graph? Explain.

Activity 1-2: The Magnitude of Gravitational Acceleration

As you saw in Lab 2, you can find a value for the average acceleration in two ways. Method 1 is to read the average value from the acceleration–time graph. Method 2 is to find the slope of the velocity–time graph.

1. For Method 1, use the **analysis and statistics features** of the software to read the average value of the acceleration during the time interval from just after the ball began falling (beginning of uniform acceleration) to just before the ball stopped falling (end of uniform acceleration).

   Average acceleration:______m/s^2

2. For Method 2, use the **fit routine** to find the mathematical relationship between velocity and time during the same time interval as in (1).

   Write below the equation you find from the fit that relates velocity (v) to time (t).

3. From the equation, what is the value of the acceleration?

   Average acceleration:______m/s^2

Question 1-4: Did the two values for the gravitational acceleration agree with each other? Should they agree with each other?

Question 1-5: What is the meaning of the other constant in the fit equation for v vs. t? What velocity of the ball does this represent?

Activity 1-3: Motion Up and Down

Prediction 1-3: Suppose that you toss a ball upward and analyze the motion as it moves up, reaches its highest point, and falls back down. Is the acceleration of the ball the same or different during the three parts of the motion—moving upward, momentarily at the highest point, and moving downward? Explain.

Prediction 1-4: Sketch on the axes below your predictions of the velocity–time and acceleration–time graphs for the entire motion of the ball from the moment it leaves your hand until just before it returns to your hand. Assume that the positive direction is upward.

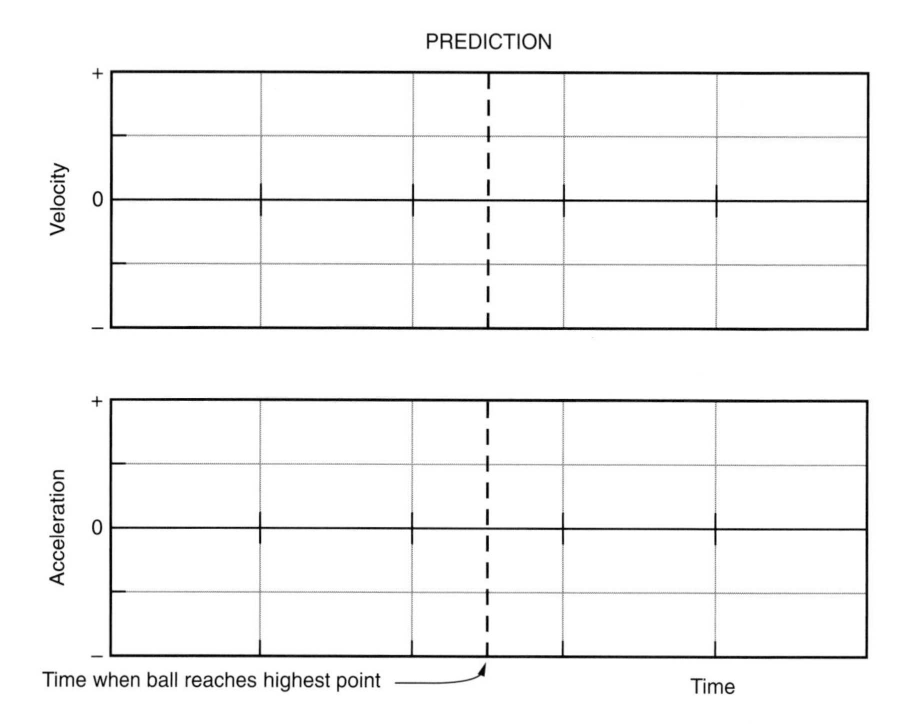

Throwing the ball upward and keeping it in the range of the motion detector is harder to do than dropping the ball. Try to throw the ball up directly under the motion detector. It may take a number of tries. *Again be sure that your body is not seen by the motion detector.*

1. The same experiment file **Falling Ball (L06A1-1)** used in Activity 1-1 should work in this activity as well. Keep the graphs from Activity 1-1 **displayed persistently on the screen.**

2. When everything is ready, **begin graphing,** and when you hear the clicking begin, toss the ball up toward the motion detector. Toss the ball as high as you can, but remember that it should never get closer than 0.5 m below the motion detector. A short quick throw will work best.

   Repeat until you get a throw where you are sure the ball went straight up and down directly below the detector.

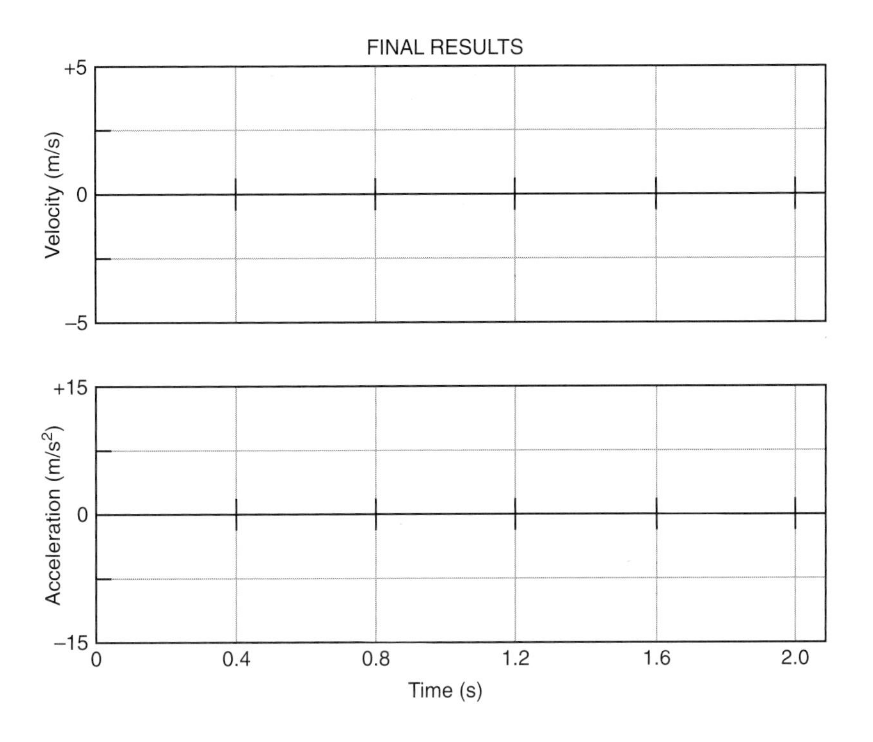

When you get a good run, sketch the portion of your graphs corresponding to the ball's up and down motion on the axes above, or **print** the graphs and affix them over the axes. Label the portions of the printed graphs that show the ball's up and down motion.

Also label with an arrow the instant in time when the ball reached its highest point.

Question 1-6: Compare the graphs to your predictions. In what ways do they differ and in what ways are they the same?

Question 1-7: What motion of a different object observed in a previous lab resulted in similar graphs to the ones for the ball thrown up in the air? Describe what was moving and precisely how it was moving.

Question 1-8: Qualitatively compare the acceleration during the three parts of the motion—on the way up, at the highest point, and on the way down. Explain your observations based on the sign of the change in velocity, as you did in Lab 2.

Question 1-9: Compare the portion of the acceleration graph when the ball was falling downward (after reaching its highest point) to the acceleration graph from Activity 1-1 where the ball was falling from rest. Are they similar? Explain why or why not.

If you have more time, do the following Extension.

Extension 1-4: Vertical Motion with Air Resistance

In this Extension you will use the motion detector to examine the motion of a paper coffee filter falling from rest. In addition to the setup in Activity 1-1 you will need

- flat-bottomed paper coffee filter—the type with a flat bottom and folds along the sides

Use the same experiment file as in Activity 1-1, **Falling Ball (L06A1-1).** If necessary, increase the time range to record the complete motion of the filter, and **adjust the velocity and acceleration axes** if necessary to display the graphs more clearly.

Be sure to keep your body out of the way of the motion detector.

Sketch the graphs on the axes that follow, or **print** and affix them over the axes.

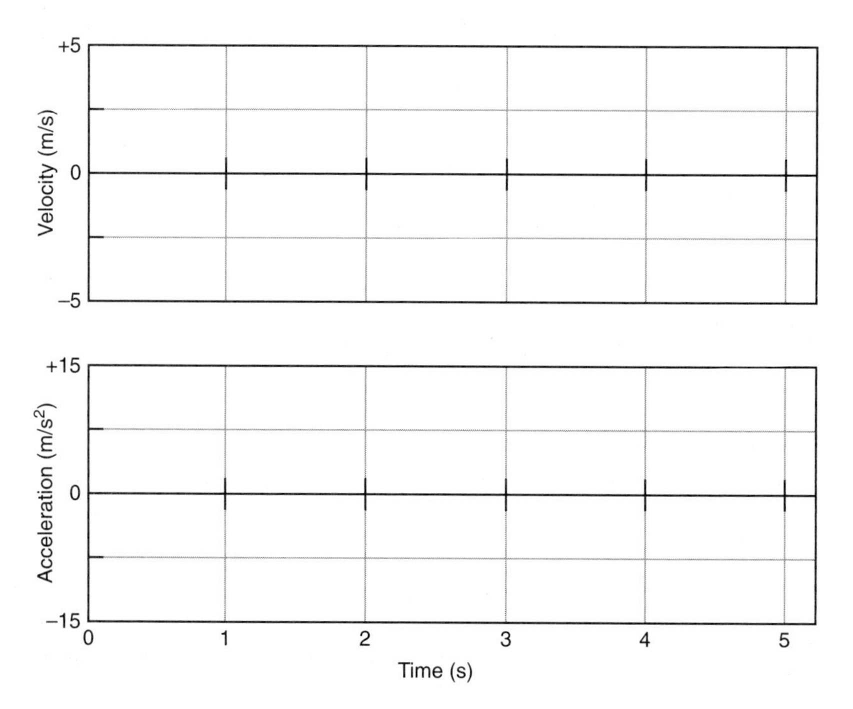

Question E1-10: Compare the graphs to those for the falling ball. Does the filter also appear to fall with a constant acceleration? If not how would you describe the motion? Is the velocity changing as the filter falls. If so, how?

Question E1-11: Based on *Newton's laws of motion,* do you think that the gravitational force is the only force acting on the filter? If there is another force, what is its direction and how does its magnitude compare to the gravitational force? Explain.

Another common motion involving the gravitational force is that of an object along an inclined ramp.

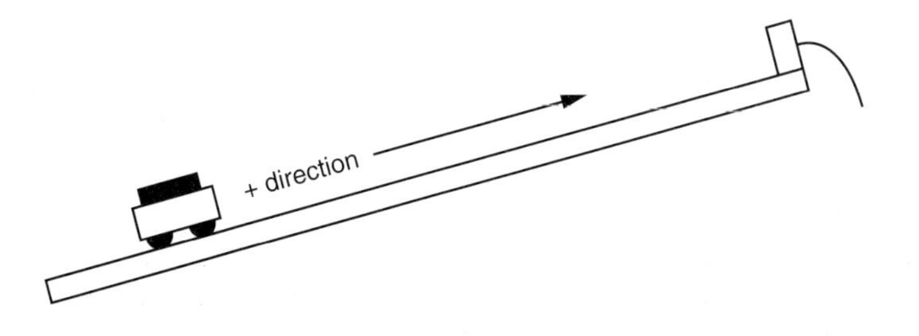

Prediction 1-5: Consider a very low-friction cart moving along an inclined ramp, as shown above. The cart is given a push up the ramp and released, and its motion is graphed using the motion detector at the top of the ramp. (The direction *toward* the motion detector is *positive*.) Sketch on the axes below your predictions of the velocity–time and acceleration–time graphs for the motion of the cart.

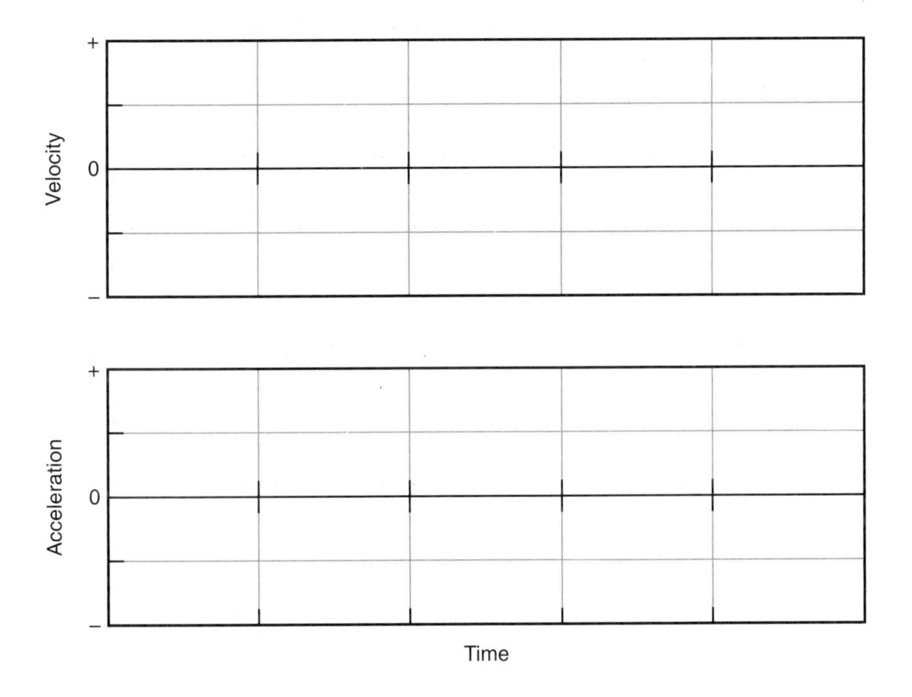

Prediction 1-6: How does the magnitude of the acceleration of the cart compare to the acceleration you determined in Activity 1-2 for a ball falling straight downward? Explain.

To test your predictions you will need in addition to the equipment used above

- low-friction cart
- smooth ramp 2–3 m in length and support to raise one end about 10 cm

Activity 1-5: Motion Along an Inclined Ramp

1. Set up the ramp with the motion detector at the high end, which has been elevated 10 cm or so. Be sure that the motion detector sees the cart over the whole length of the ramp.

2. Open the experiment file called **Inclined Ramp (L06A1-5)** to display the axes that follow. As before, the software has been set up to consider motion toward the detector positive so that *the positive direction of motion is up the ramp.*

3. If the cart has a friction pad, it should be raised out of contact with the ramp. Hold the cart at the bottom of the ramp and **begin graphing.** When you hear the clicks of the motion detector, give the cart a push up the ramp and release it.

   The graph should include all three parts of the motion—up the ramp, at its highest point, and on its way down—and the cart should never get closer than 0.5 m from the motion detector. Repeat until you have good graphs.

4. Sketch the graphs on the axes below or **print** and affix them over the axes.

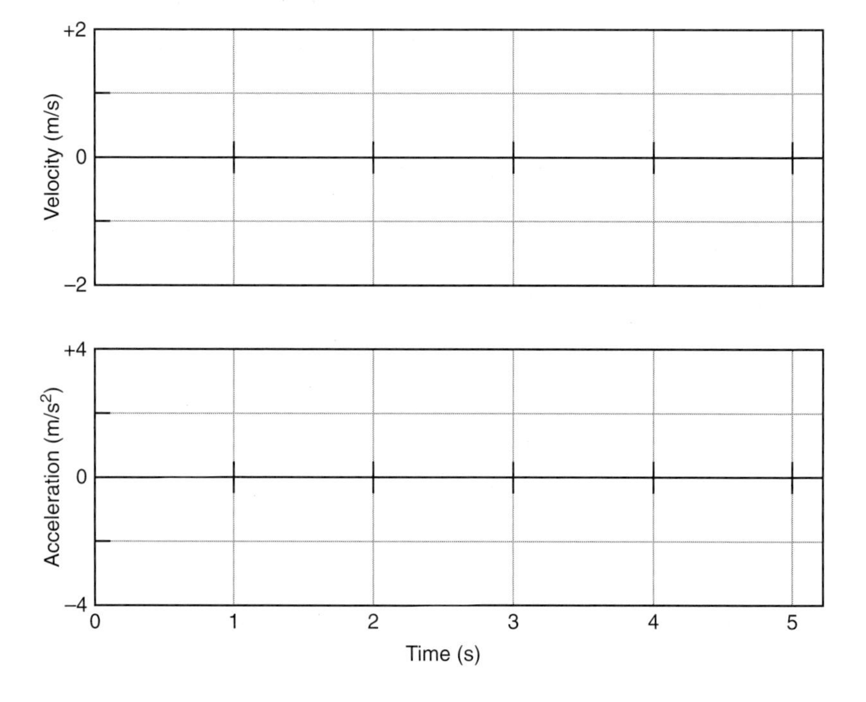

5. Use an arrow to mark the instant when the cart was at its highest point along the track on each graph.

6. Use the **analysis and statistics features** of the software to measure the average acceleration of the cart during the time interval after it was released and before it was stopped at the bottom of the ramp.

   Average acceleration: ______m/s^2

Question 1-12: Did your graphs agree with your predictions? Does this motion appear to be with a constant acceleration?

Question 1-13: How did the magnitude of the acceleration compare with that for the ball falling which you determined in Activity 1-2? Is this what you predicted?

Question 1-14: Was this motion caused by the gravitational force? Why isn't the acceleration the same as for the ball falling?

If you have more time, do the following Extension to look at this motion more quantitatively.

Extension 1-6: Analysis of Cart Moving on an Inclined Ramp

You can analyze motion of the cart on an inclined ramp more quantitatively using a fan unit.

In addition to the materials for Activity 1-5, you will need

- fan unit
- force probe

Mount the fan unit securely on the cart so that it will apply a force up the ramp. The motion detector should be at the top of the ramp, as in Activity 1-5.

Turn on the fan unit and quickly adjust the tilt of the ramp until the cart and fan unit remain at rest on the ramp. Support the end of the ramp so that it remains tilted at this angle. Turn off the fan unit.

Question E1-15: How does the fan unit keep the cart at rest on the incline? What force is it balancing? What will happen if you shut off the fan and let the cart go?

Turn the fan unit off and begin graphing, as you release the cart from the top of the ramp (further than 0.5 m away from the motion detector). Determine the average acceleration of the cart as it moves down the ramp.

Measure the mass of the cart and fan unit. Use the force probe to measure the force of the fan unit, as you did in Lab 4, Activity 2-2. Be sure to **calibrate** the force probe or **load the calibration.** Then use *Newton's second law* to calculate the acceleration the cart would have if pushed by this force.

Question E1-16: How does this calculated acceleration compare to the average acceleration you measured for the cart moving down the ramp with the fan unit off? How should it compare?

INVESTIGATION 2: WHAT IS GRAVITY?

Hey, look, no hands! The ball in Activities 1-1 to 1-3 was released from rest and fell with a *uniform negative acceleration* in the y direction without the aid of a *visible* applied force. But if *Newton's second law* holds, then the net force in the y direction should equal the mass of the ball times its acceleration. We get

$$F_y^{net} = \Sigma F_y = ma$$

where the magnitude of a is equal to g, the local gravitational field strength. Maybe a belief in *Newton's second law* can help us explain the nature of gravity mathematically. To do the next two activities about the nature of gravity you will need the following items:

- balance or electronic scale reading in grams or kilograms
- rubber ball (e.g., a racquetball)
- massive steel ball
- spring scale with a maximum reading of 5 N

Activity 2-1: Discovering Gravity

First, you should describe the nature of the force that could cause the acceleration of the ball that you observed.

1. Use the balance or electronic scale (but not the spring scale) to determine the mass of the rubber ball in kilograms, and write it in the space below.

$m =$ ______kg

Question 2-1: Suppose the ball was floating in outer space (away from the gravity of the Earth, friction, or any other influence) and that *Newton's second law* holds. Calculate the force in newtons that you would have to apply to the ball so it would move with an acceleration of the magnitude that you just observed in Activity 1-1 toward the Earth.

Question 2-2: What would the direction of the force need to be? How do you know?

Question 2-3: If *Newton's second law* is to be used in the situation where you dropped the ball in the laboratory (on the Earth) with no *visible* applied force on it, what force do you need to *invent** to make *Newton's second law* valid? Is the force constant or varying during the time the ball is falling? What is its magnitude? Its direction?

So far you have studied the motion of just one object under the influence of the gravitational force you invented (or discovered). You should have observed that the acceleration of the falling ball was constant so that the gravitational force was constant. This doesn't tell the whole story. How does the mass of a falling object affect its acceleration? Is the gravitational force constant—independent of

*If you already believe *Newton's second law* is a fundamental law of nature, then you might prefer to say you *discovered* the gravitational force. If you feel you and Newton have been constructing this law on the basis of some interplay between your minds and nature's rules then you could say you are *inventing* the idea of the gravitational force.

the mass of the falling object just the way a horizontal push or pull with your hand on a cart might be constant?

Activity 2-2: Gravitational Force and Acceleration When Different Masses Fall

Prediction 2-1: If you were to drop a massive steel ball and a not very massive rubber ball at the same time, would they fall with the same acceleration? Explain the reasons for your prediction.

1. Release the two objects from the same height at the same time, and observe when they hit the floor. Repeat several times.

Question 2-4: Did one object take significantly longer than the other to reach the floor, or did they both hit at about the same time?

Question 2-5: What do you conclude about the accelerations of the two objects—Is one significantly larger, or are they both about the same in magnitude?

2. Use a balance or an electronic scale (but not the spring scale) to determine the masses of the two objects in kilograms.

$$m_{\text{rubber}} = \underline{\qquad\quad}\text{kg}$$
$$m_{\text{steel}} = \underline{\qquad\quad}\text{kg}$$

3. Although you only made a casual qualitative observation of the objects you dropped, it turns out that in the absence of other forces such as air resistance all objects falling close to the Earth's surface have the same magnitude acceleration given by g = 9.8 m/s^2. (There are small variations from place to place and, of course, uncertainties in measurements.) Assuming both the rubber and steel balls accelerate at this same standard rate, use *Newton's second law* to calculate the magnitude of the gravitational force exerted on each one

$$F^{\text{grav}}_{\text{rubber}} = \underline{\qquad\quad}\text{ N}$$
$$F^{\text{grav}}_{\text{steel}} = \underline{\qquad\quad}\text{ N}$$

Question 2-6: If you have any object of mass m accelerating at a constant rate given by g, what is the equation that you should use to determine the gravitational force F^{grav} on it? The gravitational force, F^{grav}, is often referred to as its weight. (See the Comments that follow.)

4. Check out some weights using a spring scale that has been calibrated in newtons and fill in the table that follows.

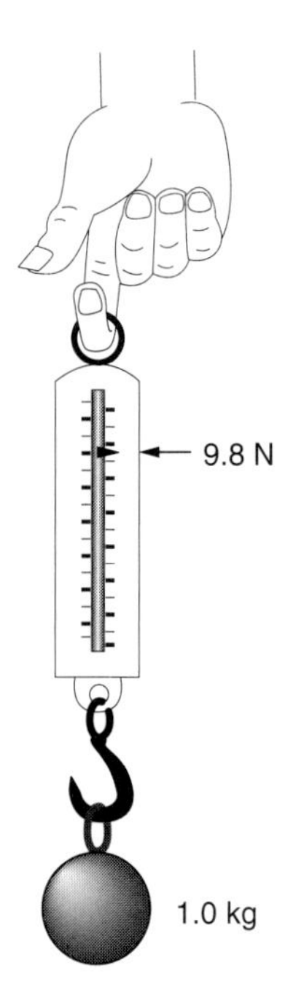

Object	Measured with balance mass (kg)	Calculated with Newton's second law F_g (N)	Measured with spring scale weight (N)
Rubber ball			
Steel ball			

Comments on Mass, Force, and Newton's Laws: *What is mass?* Philosophers of science had great debates about the true definitions of force and mass. For example, some interpret *Newton's second law* as the definition of force, with mass and acceleration being the most fundamental quantities. Others feel that force and acceleration are fundamental quantities, and that *Newton's second law* can be used to define mass.

If we assume that mass refers somehow to "amount of stuff," then we can develop an operational definition of mass for matter that is made up of particles that appear to be identical. We can assume that mass adds up and that two identical particles have twice the mass of one particle, three particles have three times the mass, and so on.

But suppose we have two objects that have different shapes and are made of different stuff, such as a small lead pellet and a silver coin. To compare their masses we can put them on a balance, and when they balance each other we say that the "force due to gravity" or the force of attraction exerted on each of them by the Earth is the same, so they must have the same mass.

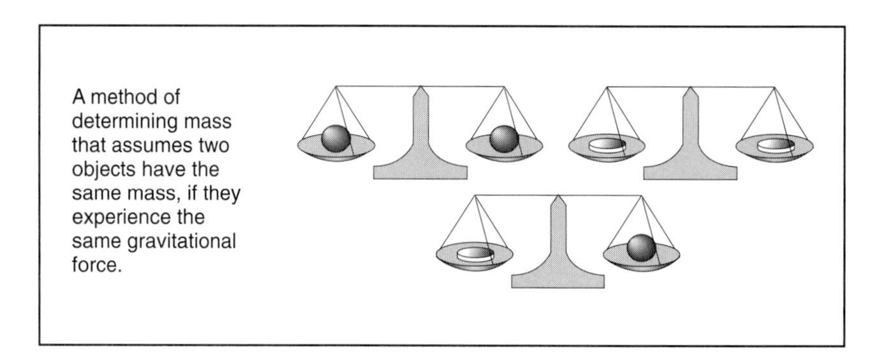

A method of determining mass that assumes two objects have the same mass, if they experience the same gravitational force.

Actually, if we balance gravitational forces as the method of determining mass, we are only determining a *gravitational mass.* Gravitational mass is proportional to the force of attraction exerted by the Earth on the mass.

Another approach to determining mass that we used in Lab 5 is to apply a constant force to an object, measure its acceleration, and calculate the mass as the ratio F/a. This method is used to determine *inertial mass.* Inertial mass is a measure of the resistance of an object to acceleration.

It is not obvious that these two definitions of mass—*gravitational* and *inertial*—should yield the same results. This equivalence is assumed in both Newton's theory of gravity and Einstein's general relativistic modifications of it. Mass can be measured with modern techniques to about 1 part in 10^{11}. Sophisticated experiments have shown that within these limits of experimental uncertainty, there is no difference between the two types of mass.

What is force? Force can be defined several ways, which happily seem to turn out to be consistent. (1) It can be defined as a push or pull and measured in terms of the stretch of a rubber band or spring, or the reading of a force probe. (2) Alternatively, the net or combined force on an object can be defined as the cause of motion. In this case we use *Newton's second law* to define net force by considering what happens to a standard mass.

Suppose we extend a spring just enough so that a mass of "exactly" one kilogram will accelerate by exactly one meter per second squared when this force is applied to it. That force is *defined* as one newton. (An apple like that which Newton is supposed to have contemplated in free fall feels a gravitational force of about one newton.) (3) Finally, we can define force in terms of the pull exerted on a mass by the Earth as determined by the stretch of a spring when a mass is hanging from it. Thus, force is both a push or a pull and the cause of motion (for a nonzero net force).

Is there a difference between mass and weight? Weight is a measure of the gravitational force F_g on a mass m, and mass is a measure of its resistance to motion. Many individuals confuse the concepts of mass and weight. Now that you understand *Newton's laws,* you should know the difference.

Activity 2-3: Mass and Weight

Test your understanding by answering the questions posed below.

Question 2-7: If mass is a measure of the amount of "stuff" in an object, is an astronaut's mass different on the moon?

Question 2-8: How can astronauts jump so high on the moon? Is the astronaut's weight different on the moon than on the Earth? Explain.

Question 2-9: If weight is a force, what is pushing or pulling on the object? How is weight related to the acceleration of gravity?

Question 2-10: When do astronauts experience weightlessness? Could they ever experience masslessness?

INVESTIGATION 3: NORMAL FORCES

In this investigation you will consider the characteristics of another "invisible" type of force—*normal force*—that must be taken into account for the application of *Newton's laws* to problems of real interest.

A book resting on a table does not move, neither does a person pushing against a wall. According to *Newton's first law,* the combined (net) force on the book and on the person's hand must be zero, since neither is moving. We have to invent a type of force to explain why books don't fall through tables and hands don't usually push through walls. Since the forces exerted by a surface always seem to act in a direction perpendicular to the surface, such forces are called *normal forces.* (Normal is a synonym for perpendicular.) Normal forces are examples of *passive* forces because they seem to act in response to *active* forces like pushes and gravitational forces. (Other forces characterized as *passive* are tension and frictional forces. These will be examined in Lab 7.)

To investigate some attributes of normal forces you will need the following apparatus:

- embroidery hoop with elastic rubber diaphragm
- embroidery hoop with stiffer (less flexible) material such as several layers of cloth
- table top
- wall
- 50- and 100-g masses

Activity 3-1: Calculating Normal Forces

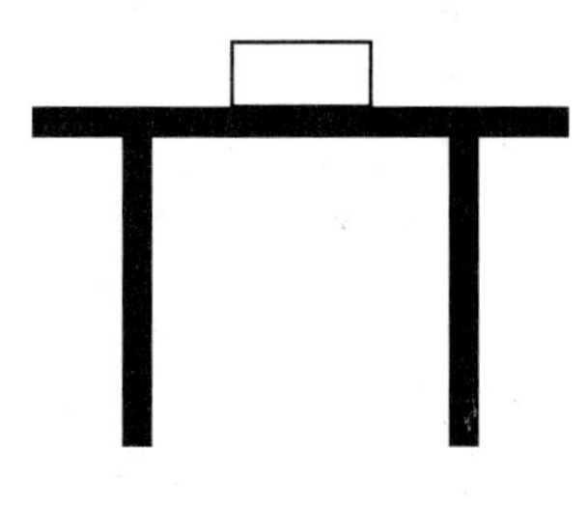

In the diagram on the right, a book is sitting on the table. The book is not moving, so the net force on it must be zero.

Question 3-1: If the mass of the book is m, what is the magnitude of the gravitational force on the book? What is the direction of this force?

Question 3-2: If the net force on the book is zero according to *Newton's first law,* then what magnitude force must the table exert on the book? What is the direction of this force?

Question 3-3: What would the book do if the table didn't exert a force on it? (Suppose that the table is suddenly removed.)

Activity 3-2: A Mechanism for Normal Forces

By applying forces perpendicular to flexible surfaces with different degrees of stiffness you can discover a mechanism for the passive normal forces that crop up in reaction to forces applied to a surface.

1. Hold the embroidery hoop with the elastic rubber vertical and press in the center perpendicular to the rubber surface. Observe what happens to the center.
2. Press harder in the center of the rubber surface and again observe what happens to the center.
3. Repeat steps 1 and 2 using the stiffer (less flexible) hoop. Try to apply pushes of about the same magnitude as in steps 1 and 2, and ob-

serve how much the surface is displaced and how hard the surface pushes back on your finger.

4. Do a similar set of investigations using the same hoops held horizontally with 50-g, then 100-g, and finally 150-g masses placed at the centers of the surfaces.

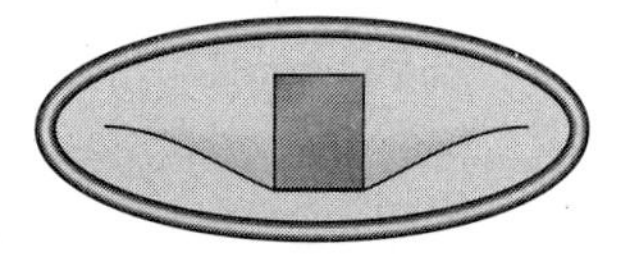

Question 3-4: When the rubber surface has a force applied to its center, does it push back? How do you know? Does the force seem to change as it is pushed harder?

Question 3-5: What happens to the center of the rubber surface when it is pushed? What happens as it is pushed harder?

Question 3-6: Did the surface of the stiffer material bend more or less than the more flexible rubber surface when the same force was applied? Explain.

Question 3-7: What mechanism do you think might explain the ability of the surfaces to react to an active force by applying a normal force?

Question 3-8: Based on your observations with the stiffer surface, what would happen to the bending of the surface if it was made up of hundreds of layers of cloth to resist a person's push or hold up a mass?

Question 3-9: Does a table or wall bend *noticeably* if an active force is applied to it? What mechanism do you propose to explain how walls and tables can exert normal forces? Do you think that the surface bends at all? Explain.

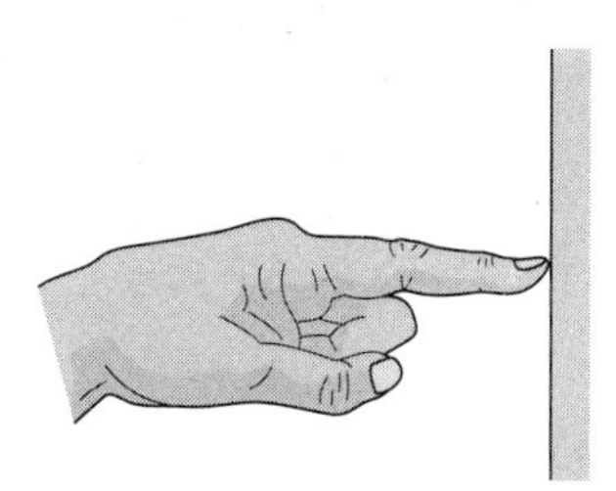

Name______________________ Date______________ Partners______________________

HOMEWORK FOR LAB 6: GRAVITATIONAL FORCES

1. Explain how a careful analysis of the motion of an object falling close to the Earth's surface leads to the conclusion that there must be a gravitational force of constant magnitude acting on the object in a direction toward the ground. (**Hint:** Carefully describe the motion, and use *Newton's laws* in your explanation.)

2. Explain how a careful analysis of the motions of two objects of different mass falling close to the Earth's surface leads to the conclusion that the gravitational force is proportional to the mass of an object. (**Hint:** Carefully describe the accelerations of the objects, and use *Newton's laws* in your explanation.)

3. You toss a coin straight up into the air. Sketch on the axes below the velocity–time and acceleration–time graphs of the coin from the instant it leaves your hand until the instant it returns to your hand. Assume that the positive y direction is upward. Indicate with arrows on your graphs the moment when the coin reaches its highest point.

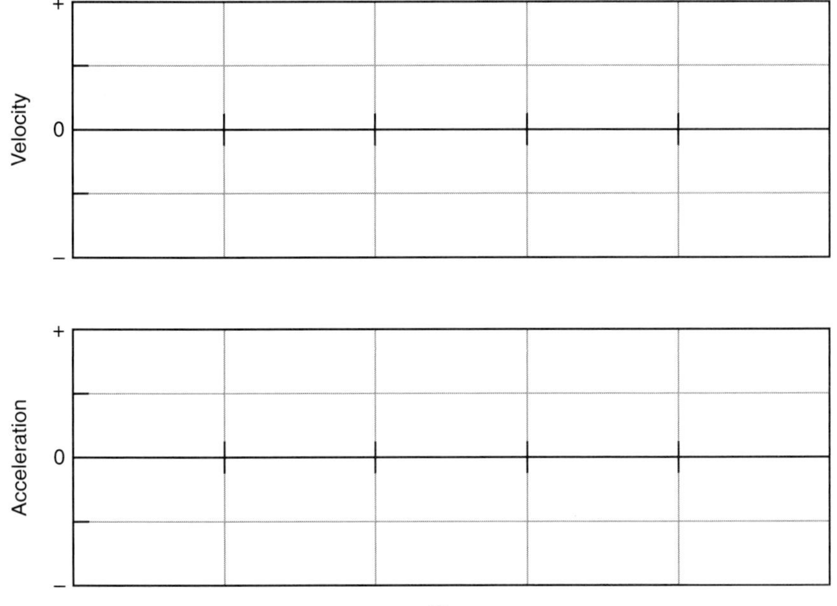

4. Based on your knowledge of the gravitational force near the surface of the Earth, and on *Newton's second law,* explain the sign and magnitude of the acceleration of the coin in Question 3 during the three portions of its motion: on the way up, at the instant when it reaches its highest point, and on the way down.

5. Based on the way the velocity is changing, explain the sign and magnitude of the acceleration of the coin in Question 3 during the three portions of its motion: on the way up, at the instant when it reaches its highest point, and on the way down.

6. Write down the mathematical relationship between velocity (v) and time (t) for a coin tossed up into the air after it leaves the thrower's hand at $t = 0$ with a velocity of $+15.0$ m/s. Assume that the positive direction is upward and the acceleration of the coin is -9.8 m/s^2.

7. What is the weight of a ball of mass 55.0 kg? What is the magnitude of the gravitational force acting on it? What force would a spring scale read if the ball were hanging from it?

8. Consider two identical balls. One is hanging on a spring scale in the laboratory, and one is hanging on a spring scale 25 km above the surface of the Earth. Which ball weighs more? Which ball has a larger mass?

9. When a book weighing 10 N is placed on a table, the table exerts a normal force of 10 N on the book. When a crate weighing 250 N is placed on the same table, the table exerts a normal force of 250 N on the crate. Explain the mechanism that allows the table to exert just the right normal force to balance the weight of the object. Is there any limit to how large the normal force can be?

Name____________________________________ Date____________________

PRE-LAB PREPARATION SHEET FOR LAB 7: PASSIVE FORCES AND NEWTON'S LAWS

(Due at the beginning of Lab 7)

Directions:
Read over Lab 7 and then answer the following questions about the procedures.

1. What is the friction pad used for in Investigation 1?

2. Just before Prediction 1-1, you adjust the friction pad on the bottom of the cart so that the cart moves at a constant velocity as the 50-g mass falls. Describe all the forces acting on the cart.

3. What is your Prediction 1-1? Will moving with a larger constant velocity mean a larger frictional force?

4. What is your Prediction 2-3? Will one person exert a larger force on the other or will the forces be equal?

5. What is the piece of wire used for in Activity 3-1?

Name________________________ Date____________ Partners________________________

Lab 7: Passive Forces and Newton's Laws

There's no way of avoiding the tension of tightrope walking.

—Anonymous circus performer

OBJECTIVES

- To incorporate frictional forces into *Newton's first and second laws of motion.*
- To explore interaction forces between objects as described by *Newton's third law of motion.*
- To explore *tension* forces and understand their origin.
- To apply *Newton's laws of motion* to mechanical systems that include tension.

OVERVIEW

In Lab 6, you had to "invent" an invisible gravitational force to save Newton's second law. Since objects near the surface of the Earth fall with a constant acceleration, you concluded by using *Newton's second law* that there must be a constant (gravitational) force acting on the object.

Finding invisible forces (forces without an obvious agent to produce them) is often hard because some of them are not *active* forces. Rather, they are *passive* forces, such as the *normal* forces that you examined in the last investigation of Lab 6, which crop up only in response to active ones. (In the case of normal forces, the active forces are ones like the push you exert on a wall or the gravitational pull on a book sitting on a table.)

Frictional and *tension* forces are other examples of passive forces. The passive nature of friction is obvious when you think of an object like a block being pulled along a rough surface. There is an applied force (active) in one direction and a frictional force in the other direction that opposes the motion. If the applied force is discontinued, the block will slow down to rest but it will not start moving in the opposite direction due to friction. This is because the frictional force is passive and stops acting as soon as the block comes to rest.

Likewise, tension forces, such as those exerted by a rope pulling on an object can exist only when there is an active force pulling on the other end of the rope.

In this lab you will use your belief in *Newton's laws of motion* to "invent" *frictional* and *tension* forces. Along the way you will examine *Newton's third law of motion.*

INVESTIGATION 1: NEWTON'S LAWS WHEN FRICTION IS PRESENT

In previous labs we have been trying hard to create situations where we could ignore the effects of friction. We have concentrated on *applied* forces involving pushes and pulls that we can see and measure directly. The time has come to take friction into account. You can make observations by applying a force directly to your force probe mounted on a cart and comparing the cart's acceleration when no friction is present to the acceleration when a friction pad under the cart drags on the track.

To make observations on the effects of friction you will need

- computer-based laboratory system
- *RealTime Physics Mechanics* experiment configuration files
- force probe
- motion detector
- spring scale with a maximum reading of 5 N
- cart with an adjustable friction pad under it
- smooth ramp or other level surface 2–3 m long
- 50- and 200-g hanging masses
- string
- wooden block
- masses to double and triple the block's mass

Activity 1-1: The Action of Friction

1. Set up the ramp, motion detector, force probe and cart. Attach the falling mass to the cart by means of a pulley and string, as shown in the figure that follows. *Be sure that the friction pad is not rubbing on the track.*

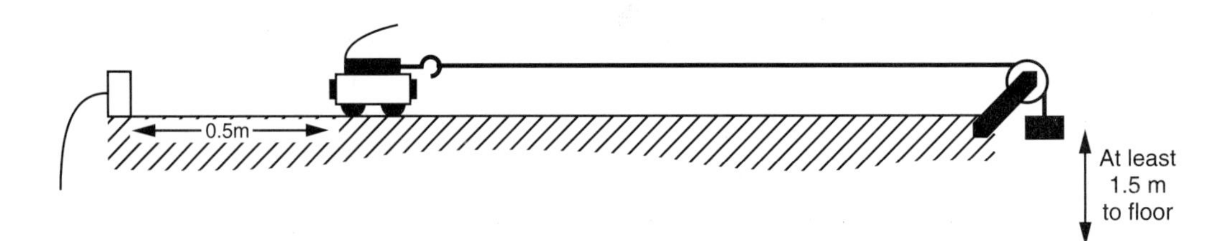

2. **Calibrate** the force probe with a force of 2.0 N applied to it with the spring scale or hanging mass or **load the calibration.** (If you are using a Hall effect force probe, you may need to adjust the spacing and **check the sensitivity.**)

3. Prepare to graph velocity, acceleration, and force by opening the experiment file called **Action of Friction (L07A1-1)** to display the axes that follow.

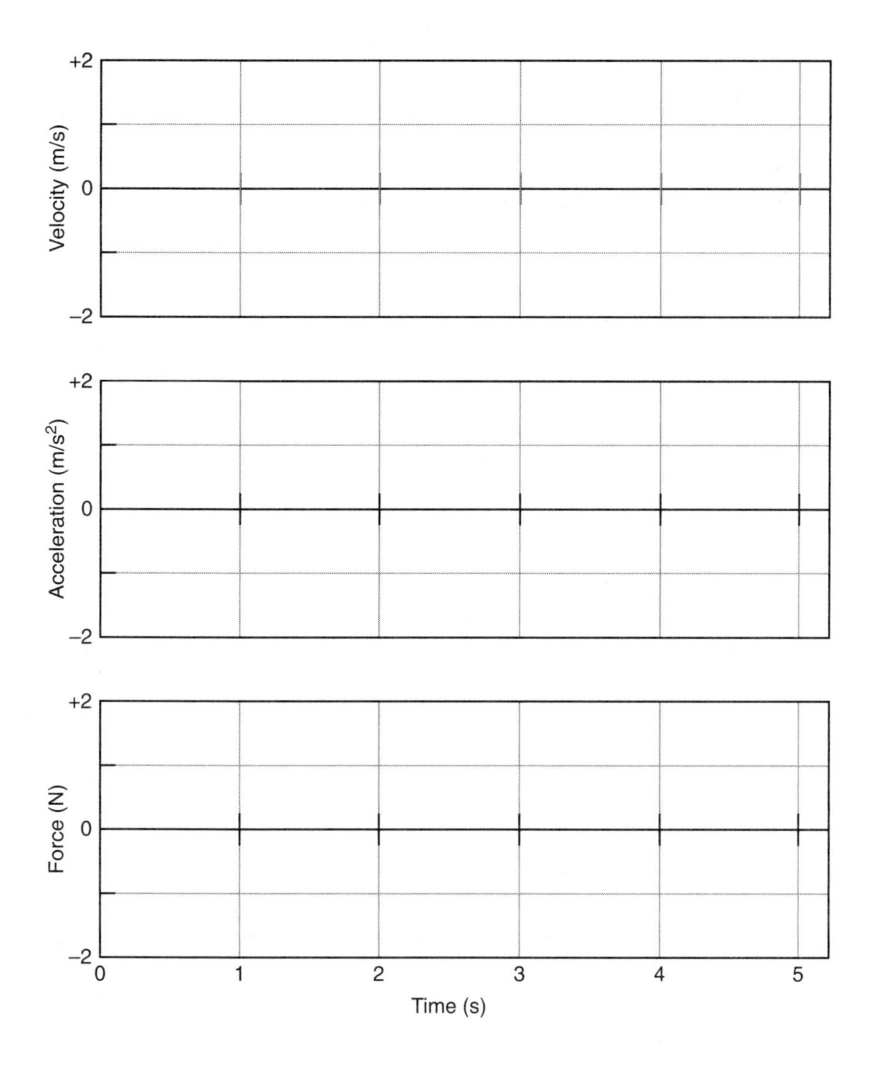

4. **Zero** the force probe with nothing pulling on it. Accelerate the cart with a hanging mass of 50 g. Move your data so that the graphs are **persistently displayed on the screen** for later comparison.

5. Sketch your graphs on the axes above, or **print** them later along with the graphs for steps 7–9 below, and affix over the axes.

6. Use the **analysis and statistics features** of your software to determine the average applied force and average acceleration of the cart.

   Record your measured values in the spaces below.

   Average applied force = ______N

   Average acceleration = ______m/s^2

7. Lower the friction pad on the bottom of the cart until it is rubbing against the track just enough to cause the cart to move at a *constant* velocity as the 50-g mass falls.

8. **Zero** the force probe with no force applied to it, and then graph the motion of the cart as before. Give the cart a little push to get it started.

9. Sketch your graphs on the same axes above with dashed lines, or **print** the graphs and affix them over the axes.

10. Find the average acceleration and the average force applied by the string on the cart, and record the values in the spaces below.

Average force applied to cart = ______N

Average acceleration = ______m/s^2

Question 1-1: If the cart can move at *constant* velocity with a force applied to it by the string, is *Newton's first law* violated? What should the *combined (net)* force on the cart be in this case if *Newton's first law* holds? Explain.

Question 1-2: If *Newton's first law* correctly describes the motion of the cart at a *constant* velocity, describe the frictional force that must be combined with the force applied by the string. Does this frictional force act in the same direction or the opposite direction as the force applied by the string? What is its magnitude? Explain.

Prediction 1-1: Suppose that the cart moves with a larger constant velocity. Will the frictional force be different from before? Explain.

11. Test your prediction. Repeat steps 7–9, only this time give the cart a larger push to get it started, so that it will move with a significantly larger constant velocity.

    Find the average acceleration and the average force applied by the string on the cart, and record the values in the spaces below.

Average force applied to cart = ______N

Average acceleration = ______m/s^2

Question 1-3: Does the magnitude of the frictional force seem to depend on the velocity of the cart? Explain.

12. Raise the friction pad a little bit so that it rubs the track *more lightly than before.* **Zero** the force probe with nothing pulling on it and again graph as you accelerate the cart with the 50-g mass falling. Record the average force applied to the cart by the string and the average acceleration of the cart in the spaces below.

Average force applied to cart = ______N

Average acceleration = ______m/s^2

Question 1-4: You should have noted that the acceleration of the cart is noticeably less than that which you observed in steps 4–6 (when the friction pad was not in contact with the ramp). Is *Newton's second law* violated? Can you *invent* a *frictional force* as in Question 1-2 to combine with the force applied by the string so that *Newton's second law* correctly describes the motion? Explain.

Question 1-5: Based on the measured acceleration and the mass of the cart, what should the magnitude of the *combined (net)* force on the cart be if *Newton's second law* is correct? Calculate the magnitude and direction of the new frictional force caused by the dragging pad if *Newton's second law* is correct.

If you have time, do the following Extensions to learn more about frictional forces.

Extension 1-2: Static and Kinetic Frictional Forces

You have seen in Activity 1-1 that when the surface of the friction pad is in contact with the surface of the track, the track exerts a frictional force on the pad. If the frictional force is equal in magnitude to the applied force, then according to Newton's first law the cart must either remain at rest or move with a constant velocity.

The frictional forces when two objects are sliding along each other are called *kinetic* frictional forces. If the objects are not sliding along each other, then the frictional forces are called *static* frictional forces.

In this Extension you will examine whether the frictional force is different when an object is at rest (*static* friction) than when it is sliding along a track (*kinetic* friction).

Prediction E1-2: A block is sitting on a table, as shown below. You pull on a string attached to the block, and the block doesn't move because the frictional force opposes your pull. You pull harder, and eventually the block begins to slide along the table top. How does the frictional force *just before* the block started sliding compare to the frictional force when the block is actually sliding? Explain.

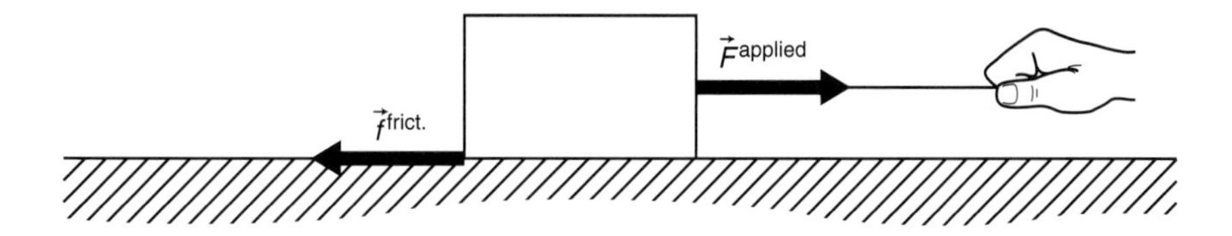

You can test your prediction by mounting the force probe on a wooden block, and pulling the block along your ramp.

Set up the block, string, force probe, motion detector, and ramp as shown below. The force probe should be taped securely to the top of the block. If you are planning to do Extension 1-3, be sure that it is possible to add masses to the top of the block to double and triple the mass of the block and force probe.

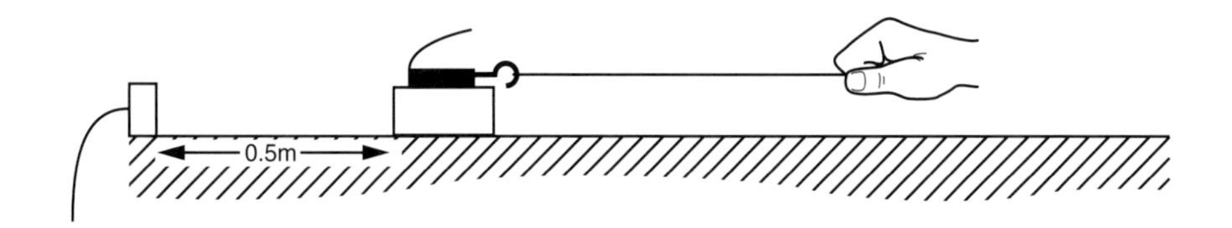

Measure the mass of the block and force probe: ______kg

Prepare to graph velocity and force by opening the experiment file called **Static and Kinetic Frictional Force (L07E1-2)** to display the axes that follow.

Calibrate the force probe with a force of about 2.0 N using the spring scale or a hanging mass or **load the calibration** if it hasn't already been calibrated. (If you are using a Hall effect force probe, you may need to adjust the spacing and **check the sensitivity.**)

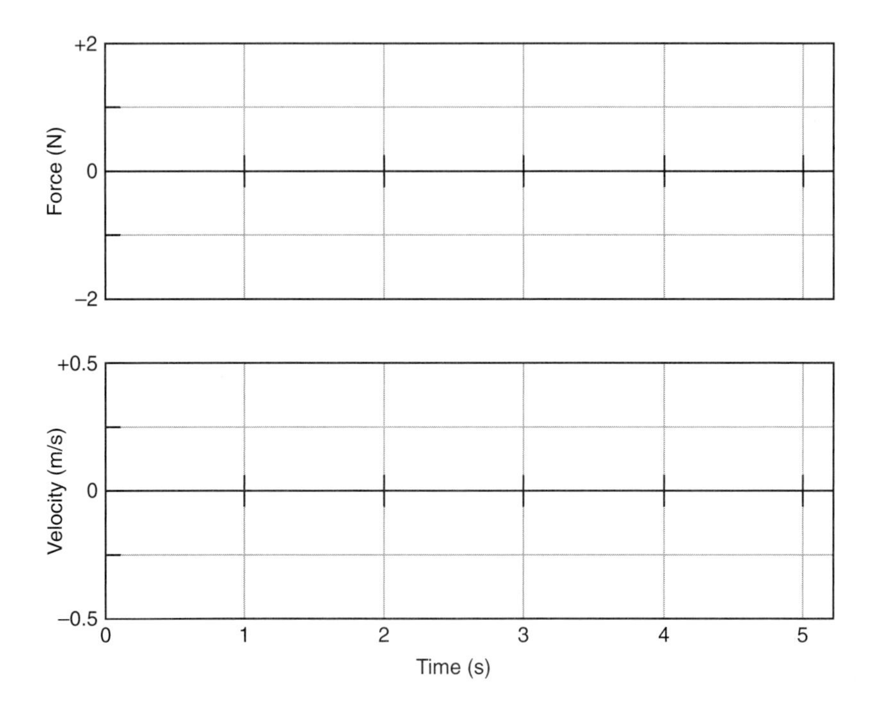

Zero the force probe with nothing pulling on it. **Begin graphing** with the string loose, then gradually pull *very gently* on the force probe with the string and increase the force very slowly. *Be sure that you pull horizontally—not at an angle up or down.* When the block begins to move, pull only hard enough to keep it moving with a small velocity which is as constant (steady) as possible.

Move your data so that the graphs are **displayed persistently on the screen** for comparison with those in Extension 1-3.

Sketch your graphs using solid lines on the axes above, or **print** and affix over the axes.

Question E1-6: Use an arrow to mark at the time on your force graph when the block just began to slide. How do you know when this time is?

Question E1-7: What happens to the frictional force *just as* the block begins to slide? Does this agree with your prediction? Is there a difference in magnitude between the *static* frictional force just before an object begins to slide and the *kinetic* frictional force when it is sliding?

Extension 1-3: Frictional Force and Normal Force

In Activity 1-1 you saw that the frictional force is nearly the same regardless of how fast the object moves. In the last extension you observed the difference between *static* and *kinetic* frictional forces. In this extension you will examine if frictional forces depend on the force between the surface and the object—the *normal* force.

Comment: All of the forces exerted on a block being pulled on a table top or ramp are shown in the diagram below. As you have seen in Lab 6, surfaces like a wall or a table top can exert a force perpendicular to the surface called a *normal* force. In the case of the block on the table, the table exerts a *normal* force upward on the block, and the Earth exerts a *gravitational* force (the weight of the block) downward on the block.

Since you know that the block doesn't move up or down as it sits at rest or slides along the table's surface, the combined (net) vertical force must be zero according to *Newton's first law*. So the magnitude of the normal force must just equal the magnitude of the gravitational force.

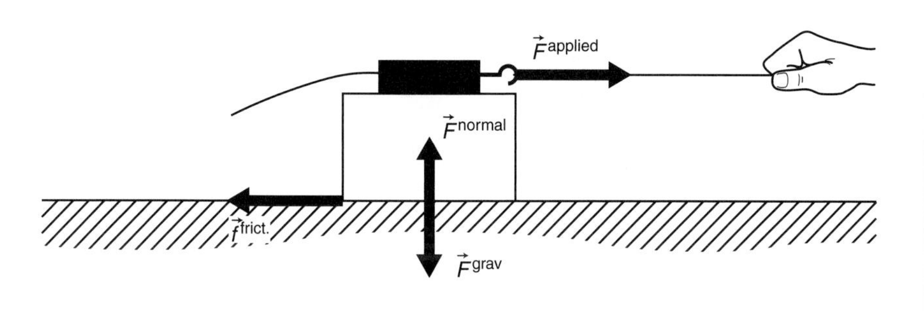

Using the symbols in the diagram, we get

$$\vec{F}^{\text{grav}} = \vec{F}^{\text{normal}} = mg$$

where m is the mass of the block and force probe, and g is the magnitude of the local gravitational field strength.

Prediction E1-3: If you increase the normal force exerted by the table on the block (by increasing the weight of the block) what will happen to the frictional force? Explain.

Test your prediction. Use the same setup and experiment file, **Static and Kinetic Frictional Force (L7E1-2)** as in Extension 1-2.

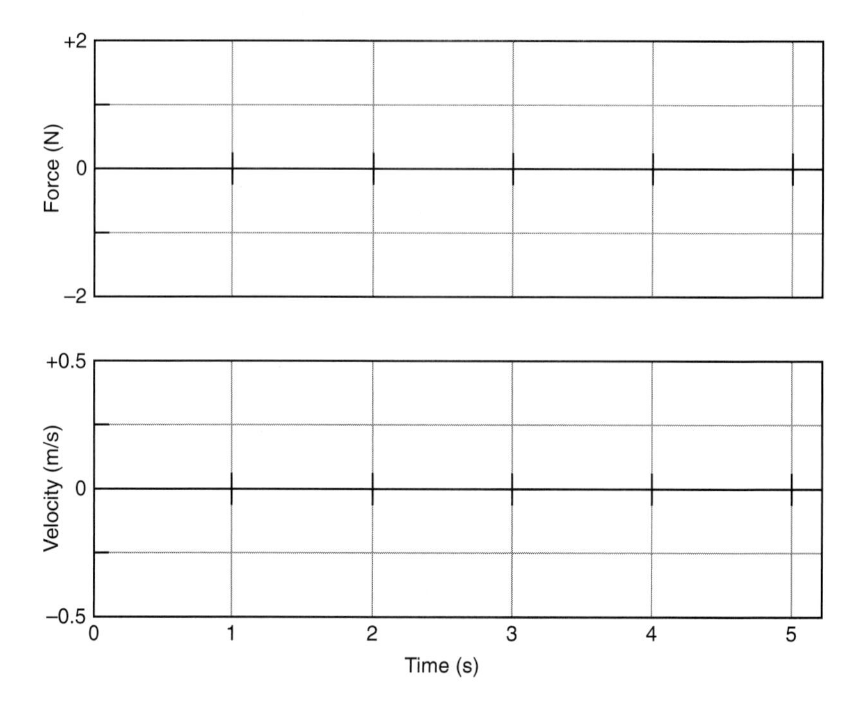

Fasten enough masses *securely* on top of the block to double its mass (including the force probe).

Record the new mass: ______kg

Calibrate the force probe with a force of 2.0 N applied to it with the spring scale or a hanging mass or **load the calibration** if it hasn't already been calibrated. (If you are using a Hall effect force probe, you may need to adjust the spacing and **check the sensitivity.**)

Zero the force probe with nothing pulling on it. As in Extension 1-2, **begin graphing** with the string loose, then gradually pull *very gently* on the force probe with the string, and increase the force very slowly. *Be sure that you pull horizontally—not at an angle up or down.* When the block begins to move, pull only hard enough to keep it moving with a small velocity that is as constant (steady) as possible.

Sketch both sets of graphs on the axes above using solid lines for the smaller mass block and dashed lines for the more massive block, or **print** the graphs and affix them over the axes.

Question E1-8: Compare the two sets of graphs. In what ways are they similar, and in what ways are they different? Are the frictional forces the same or different?

Use the **analysis and statistics features** of the software to measure the average frictional forces during the time intervals when the block was moving with a constant velocity. Record the frictional forces in the first two rows of Table E1-3 along with the normal forces calculated from the masses.

Table E1-3

	Mass of block (kg)	Normal force (N)	Average kinetic frictional force (N)
Block and force probe alone			
Doubled mass			
Tripled mass			

Graph one more time, this time with a block that has three times the mass. Be sure to **zero** the force probe before graphing. Once again pull so as to move the block with the same constant velocity. Record in Table E1-3 the mass, the normal force, and the average frictional force.

Questions E1-9: How does the kinetic frictional force appear to vary as the normal force is increased? Try to state a mathematical relationship.

INVESTIGATION 2: NEWTON'S THIRD LAW

All individual forces on an object can be traced to an interaction between it and another object. For example, we believe that while the falling ball is experiencing a gravitational force exerted by the Earth on it, the ball is exerting a force back on the Earth. In this investigation we want to compare the forces exerted by interacting objects on each other. What factors might determine the forces between the objects? Is there a general law that relates these forces?

We will begin our study of interaction forces by examining the forces each person exerts on the other in a tug-of-war. Let's start with a couple of predictions.

Prediction 2-1: Suppose that you have a tug-of-war with someone who is the same size and weight as you. You both pull as hard as you can, and it is a standoff. One of you might move a little in one direction or the other, but mostly you are both at rest.

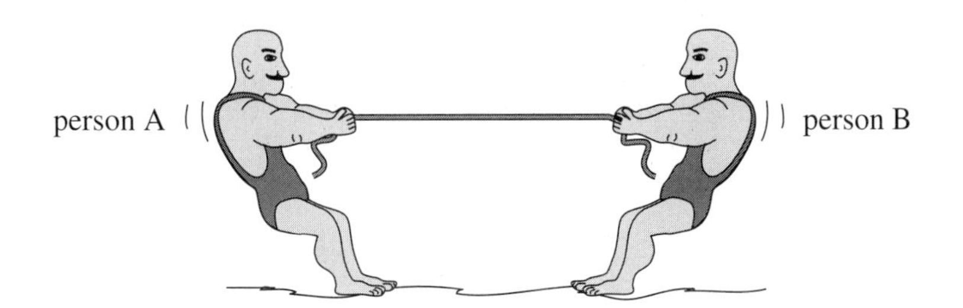

Predict the relative magnitudes of the forces between person A and person B. Place a check next to your prediction!

______Person A exerts a larger force on person B.

______The people exert the same size force on each other.

______Person B exerts a larger force on person A.

Prediction 2-2: Suppose now that you have a tug-of-war with someone who is much smaller and lighter than you. As before, you both pull as hard as you can, and it is a stand-off. One of you might move a little in one direction or the other, but mostly you are both at rest.

Predict the relative magnitudes of the forces between person A and person B. Place a check next to your prediction!

______Person A exerts a larger force on person B.

______The people exert the same size force on each other.

______Person B exerts a larger force on person A.

Prediction 2-3: Suppose now that you have a tug-of-war with someone who is much smaller and lighter than you. This time the lighter person is on a skateboard, and with some effort you are able to pull him or her along the floor.

Predict the relative magnitudes of the forces between person A and person B. Place a check next to your prediction!

______Person A exerts a larger force on person B.

______The people exert the same size force on each other.

______Person B exerts a larger force on person A.

To test your predictions you will need the following:

- computer-based laboratory system
- *RealTime Physics Mechanics* experiment configuration files
- two force probes
- two 1-kg masses to calibrate the force probes
- skateboard or pair of roller skates
- string

Activity 2-1: Interaction Forces in a Tug-of-War

1. Open the experiment file called **Tug-of-War (L07A2-1)** to display the axes that follow. The software will then be set up to measure the force applied to each probe with a data collection rate of 20 points per second.
2. **Calibrate** both force probes with a force of 9.8 N applied using hanging 1.0-kg masses or **load the calibration.** (If you are using Hall effect force probes, you may need to adjust the spacing and **check the sensitivity.**)
3. Since the force probes will be pulling in opposite directions in the tug-of-war, you should **reverse the sign** of one of them.
4. When you are ready to start, **zero** both of the force probes. Then hook a short loop of string between them, **begin graphing,** and begin *a gentle* tug-of-war. Pull back and forth while watching the graphs. *Do not pull too hard, since this might damage the force probes.*
5. Repeat with different people pulling on each side.
6. Sketch one set of graphs on the axes below, or **print** the graphs and affix them over the axes.

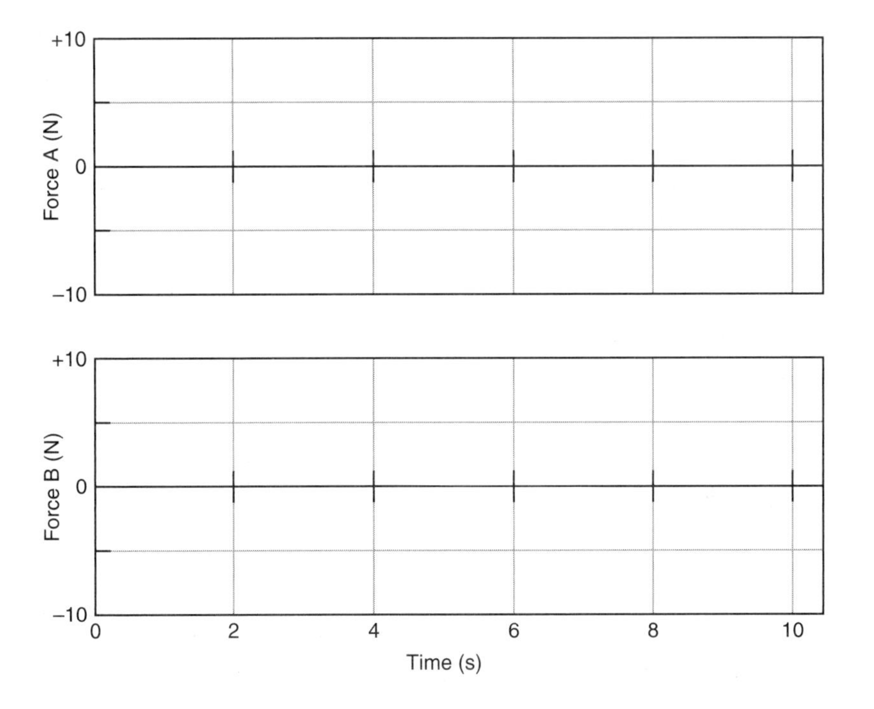

Question 2-1: How did the two pulls compare to each other? Was one significantly different from the other? How did your observations compare to your predictions?

Activity 2-2: Interaction Forces Pulling Someone Along

In this activity you will test your prediction about the interaction forces when you are pulling someone on roller skates or a skateboard along the floor.

1. You may use the same experiment file—**Tug-of-War (L07A2-1)**—as in the last activity. Be sure that the force probes have been calibrated with 9.8-N pulls. *As before, do not pull too hard, since this might damage the force probes.*
2. The person holding force probe B (person B) should be on a skateboard or wearing roller skates. **Zero** both force probes with nothing pulling on them just before taking measurements. Then hook them together, **begin graphing,** and have person A begin pulling, softly at first, then slowly increase the force until person B begins to move along the floor.
3. Sketch your graphs on the axes below, or **print** the graphs and affix them over the axes.

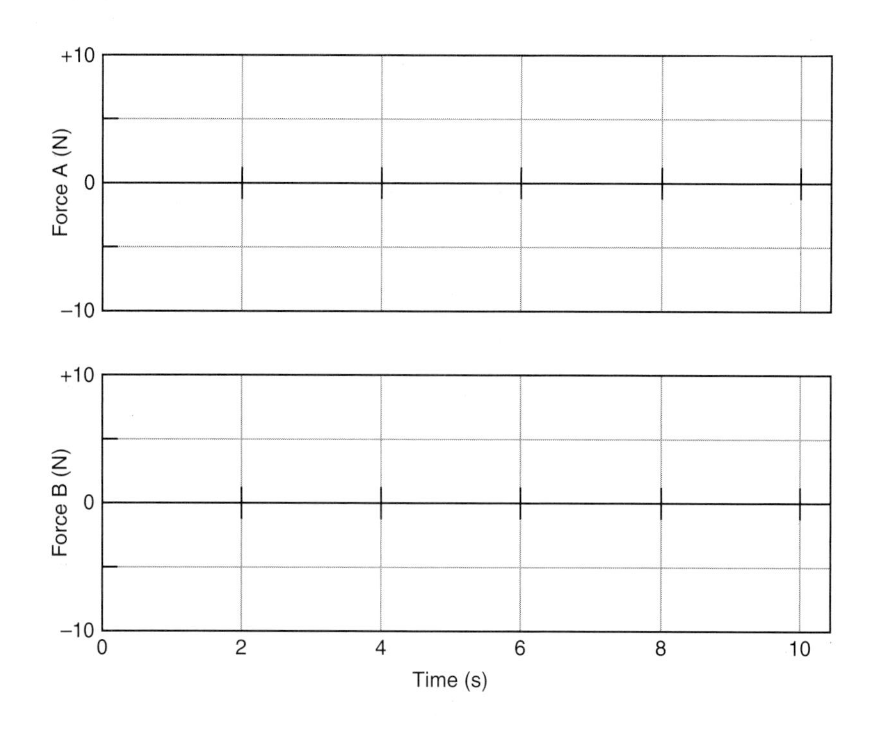

Question 2-2: How did the two pulls compare to each other? Was one significantly different from the other? How did your observations compare to your predictions?

> **Comment:** The fundamental law governing interaction forces between objects is *Newton's third law*. In contemporary English, *Newton's third law* can be stated: *If one object exerts a force on a second object, then the second object exerts a force back on the first object that is equal in magnitude and opposite in direction to that exerted on it by the first object.*

Question 2-3: Are your observations in Activities 2-1 and 2-2 consistent with *Newton's third law* of motion? Explain.

Question 2-4: When you pull on an object with a force probe, does the probe measure the force it exerts on the object or the force exerted on the probe by the object? According to *Newton's third law* does this distinction have any meaning? Explain.

> **Comment:** Newton actually formulated the *third law* by studying the interaction forces between objects when they collide. It is difficult to fully understand the significance of this law without first studying collisions, as you will in Lab 9.

INVESTIGATION 3: TENSION FORCES

When you pull on a rope attached to a crate, your pull is somehow transmitted down the rope to the crate. *Tension* is the name given to forces transmitted along stretched strings, ropes, rubber bands, springs, and wires.

Is the whole force you apply transmitted to the crate or is the pull at the other end larger or smaller? Does it matter how long the rope is? *How* is the force "magically" transmitted along the rope? These are some of the questions you will examine in this investigation.

Obviously, the rope by itself is unable to exert a force on the crate if you are not pulling on the other end. Thus, tension forces are *passive* just like *frictional* and *normal* forces. They act only in response to an active force like your pull.

Before you begin, examine your knowledge of tension forces by making the following predictions.

Prediction 3-1: If you apply a force to the end of a rope as in the picture above, is the whole force transmitted to the crate or is the force at the crate smaller or larger than your pull?

Prediction 3-2: If the rope is longer, will the force applied to the crate be larger, smaller, or the same as with the shorter rope?

Prediction 3-3: Suppose that instead of a rope, you use a bungee cord or large rubber band. Will the force applied to the crate be larger, smaller, or the same as with the rope? Suppose that you use a strong wire cable instead of a rope?

To test your predictions you will need the following:

- computer-based laboratory system
- *RealTime Physics Mechanics* experiment configuration files
- two force probes
- heavy ring stand or table clamp and rod, clamp for force probe
- two 1-kg masses to calibrate the force probes
- rubber band
- long and short pieces of string
- piece of wire the same length as the shorter string

Activity 3-1: Mechanism of Tension Forces

1. Open the experiment file called **Tension Forces (L07A3-1)** to display the axes that follow.

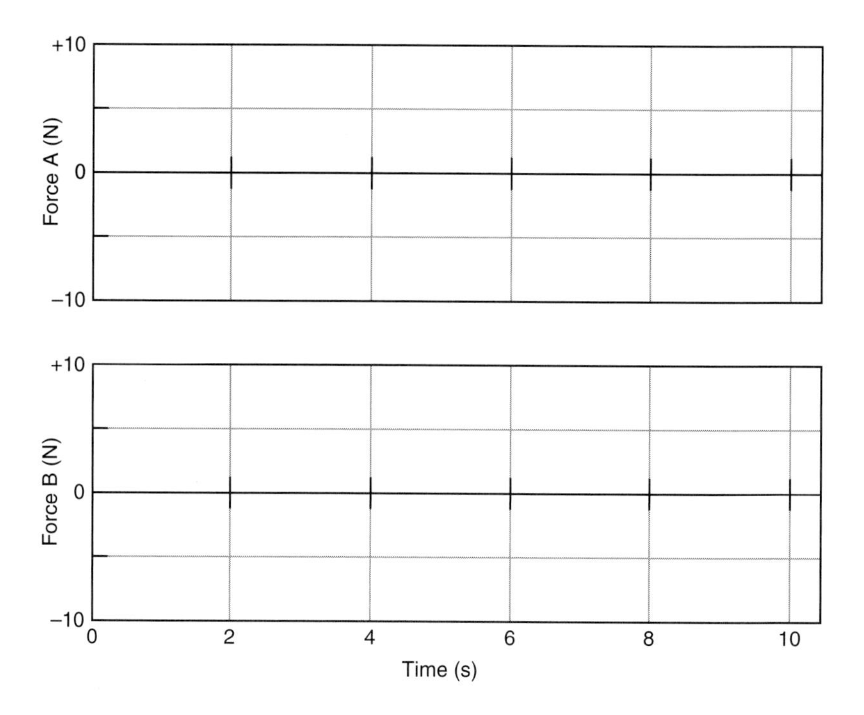

2. **Calibrate** both force probes with 1.0-kg masses (9.8 N), or **load the calibration** if you haven't already done so. (If you are using a Hall effect force probe, you may need to adjust the spacing and **check the sensitivity.**)
3. Attach force probe A horizontally to the ring stand or table clamp and rod so that it won't move when pulled.
4. Place a rubber band between the force probes.
5. **Zero** both force probes with the rubber band hanging loosely. **Begin graphing,** and pull softly at first on force probe B, then harder, and then vary the applied force. *Be sure not to pull too hard.*

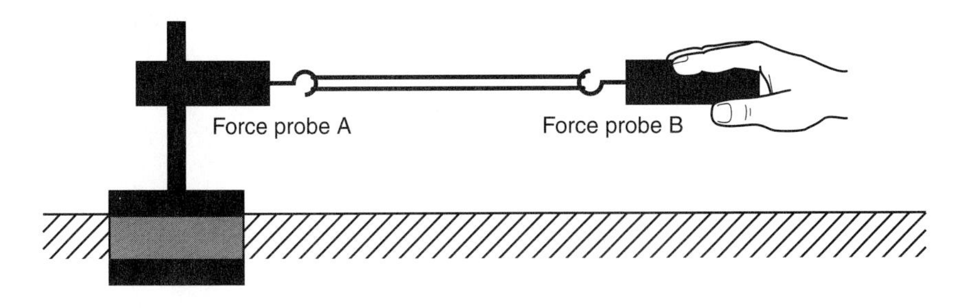

6. Sketch the graphs on the previous axes, or **print** and affix them over the axes.

Question 3-1: Based on the readings of the two force probes, when you pull on one end of the rubber band, is the force transmitted down to the other end? Explain.

Question 3-2: As you increase the force applied to the rubber band, what happens to the length of the rubber band? Propose a mechanism based on these observations to explain how the force is transmitted down the rubber band from force probe B to force probe A.

Question 3-3: Indicate with arrows on the diagram above the directions of the forces exerted by the rubber band on force probe A and force probe B.

7. Make loops at both ends of the short piece of string, and replace the rubber band with it. Repeat step 5.
8. Repeat, this time using the longer string. Be sure to **zero** both force probes before your measurements.

Question 3-4: Based on the readings of the two force probes, when you pull on one end of the string, is the force transmitted undiminished down to the other end? Does it matter how long the string is? Explain.

Question 3-5: Did the string stretch at all when you pulled on it? Can you propose a mechanism for the transmission of the force along the string?

9. Repeat, this time using the wire. Be sure to **zero** both force probes before your measurements.

Question 3-6: Based on the readings of the two force probes, when you pull on one end of the wire, is the force transmitted undiminished down to the other end? Did you observe any stretch of the wire? Do you think that the wire may have stretched a little, even though you couldn't observe the stretch with your eyes? Explain.

Prediction 3-4: What happens when a string is hung around a pulley? Is the tension force still transmitted fully from one end of the string to the other?

To test your prediction, in addition to the equipment listed above you will need

- low-friction pulley
- low-friction cart
- smooth ramp or other level surface 2–3 m long

Activity 3-2: Tension When a String Changes Direction

1. Attach force probe A securely to the cart, and set up the cart, track, pulley, string, and force probe B.

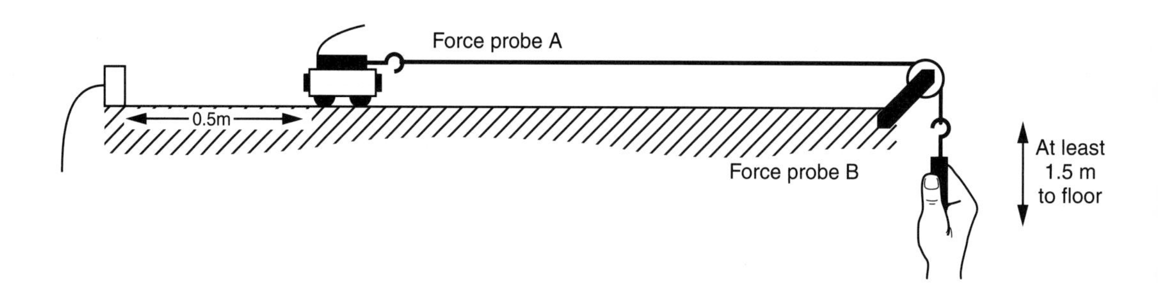

2. **Calibrate** both force probes with a hanging mass of 1.0 kg (9.8 N) or **load the calibration** if they haven't already been calibrated. (If you are using a Hall effect force probe, you may need to adjust the spacing and **check the sensitivity.**)
3. The experiment file **Tension Forces (L07A3-1)** is also appropriate for this activity.

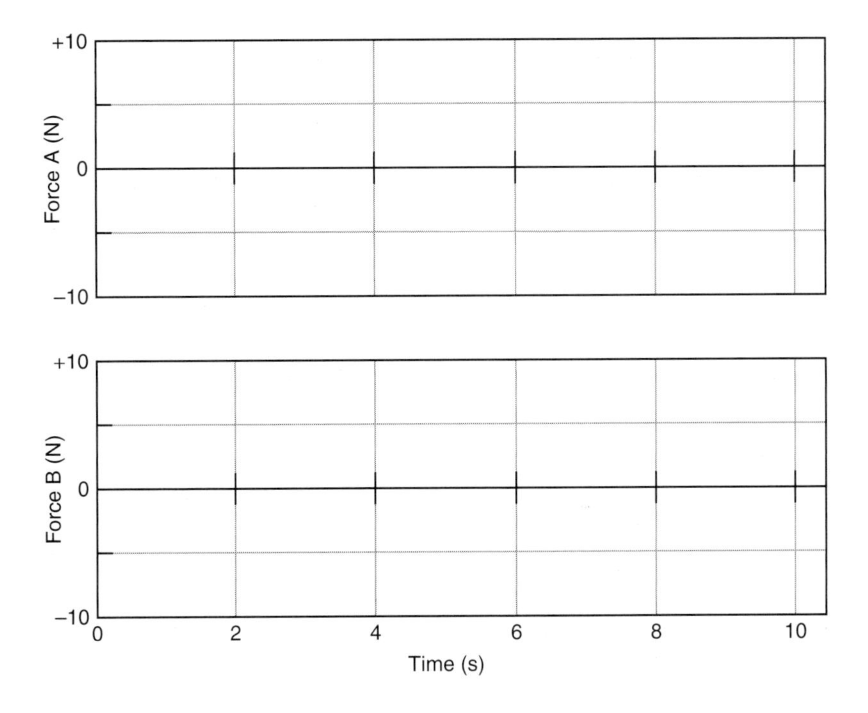

4. **Zero** both force probes with the string loose. Then **begin graphing** while pulling on force probe B and holding the cart to keep it from moving. Pull softly at first, then harder, and then alternately soft and hard. Do not exceed 10 N.
5. Sketch your graphs on the axes above, or **print** and affix them over the axes.

Question 3-7: Based on your observations of the readings of the two force probes, was the pull you exerted on force probe B transmitted undiminished to force probe A even though it went through a right angle bend? How does this compare to your prediction? Explain.

If you have additional time, do the following Extension.

Extension 3-3: Tension and Newton's Second Law

Prediction E3-5: Suppose that in place of force probe B you hang a 200-g mass from the end of the string. With the cart at rest, what value do you expect force probe A to read? Be quantitative. Explain.

Prediction E3-6: Suppose that you hang the same 200-g mass but release the cart and let it accelerate as the mass falls. Do you expect force probe A to read the same force or a larger or smaller force as the cart accelerates? Explain.

To test your predictions you will need the following in addition to the equipment for Activity 3-2:

- 200- and 500-g masses
- motion software
- motion detector

1. Use the same setup as for Activity 3-2, only replace force probe 2 with a hanging mass and set up the motion detector.
2. Determine the mass of the cart and force probe: ______kg
3. Open the experiment file called **Tension and N2 (L07E3-3)** to display the axes shown below.

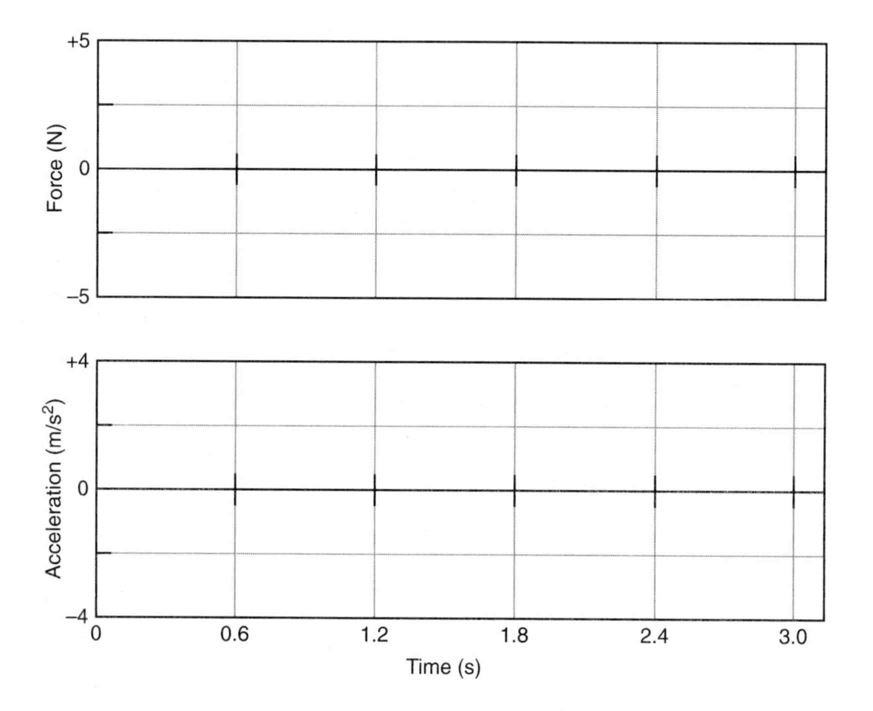

4. **Calibrate** the force probe with a 500-g mass hung over the pulley from the end of the string or **load the calibration.** (If you are using a Hall effect force probe, you may need to adjust the spacing and **check the sensitivity.**)
5. Be sure that the motion detector can see the cart all the way to the end of the track.
6. Hang the 200-g mass from the end of the string, and hold the cart at rest at least 0.5 m away from the motion detector.
7. Just before you are ready to start **zero** the force probe with the string loose. Let the 200-g mass hang without swinging and then **begin graphing.** Hold the cart for the first second after you hear the clicks of the motion detector and then release it. *Be sure to stop the cart before it comes to the end of the track.*
8. Sketch your graphs on the axes above, or **print** and affix them over the axes. Use arrows on your force and acceleration graphs to indicate the time when the cart was released.
9. Use the **analysis and statistics features** of the software to measure the average value of the tension measured by the force probe before the cart was re-

leased and while it was accelerating. Also measure the average value of the acceleration during the time interval while the cart had a fairly constant acceleration.

Average tension with cart at rest: ______N

Average tension with cart accelerating: ______N

Average acceleration: ______m/s^2

Question E3-8: How did the tension *while the cart was at rest* compare to the weight of the hanging mass? (**Hint:** Remember that *weight* = mg.) Did this agree with your prediction? Was the force applied to the end of the string by the hanging mass transmitted undiminished to the cart, as you observed in the previous activity?

Question E3-9: Did anything happen to the tension after the cart was released? How did the value of the tension *as the cart was accelerating* compare to the weight of the hanging mass? Did this agree with your prediction?

Comment: When the cart is released, the cart and hanging mass must both always move with the same velocity since they are connected by the string as a single system. Thus, the cart and mass must have the same acceleration.

According to *Newton's second law,* the combined (net) force on the cart must equal its mass times the acceleration, and the combined (net) force on the hanging mass must equal its mass times the same acceleration.

Question E3-10: Use *Newton's second law* to explain your answer to Question E3-9. Why must the tension in the string be smaller than the weight of the hanging mass if the hanging mass is to accelerate downward?

Name________________________ Date____________ Partners________________________

Homework for Lab 7: Passive Forces and Newton's Laws

1. A 1.0-N weight is hanging at rest as shown. In each case, write in the magnitude of the unknown force.

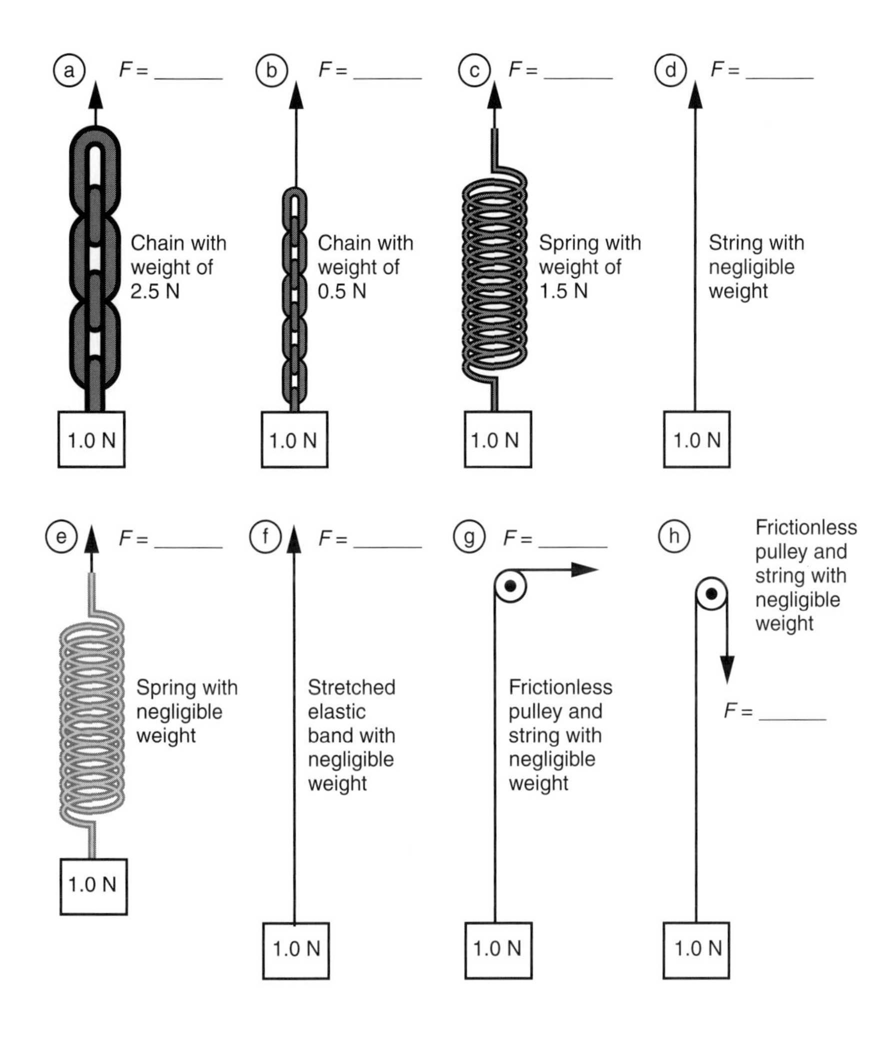

Explain the reasoning for each of your answers.

a.

b.

c.

d.

e.

f.

g.

h.

2. Explain how the tension force gets transmitted along from one end of a string to the other. Does the amount of force that gets transmitted depend on how elastic or stretchable the string is? Explain.

Questions 3–5 refer to the block on a flat surface shown below. A force F is applied to the block as shown. With an applied force of 1.5 N, the block moves with a constant velocity.

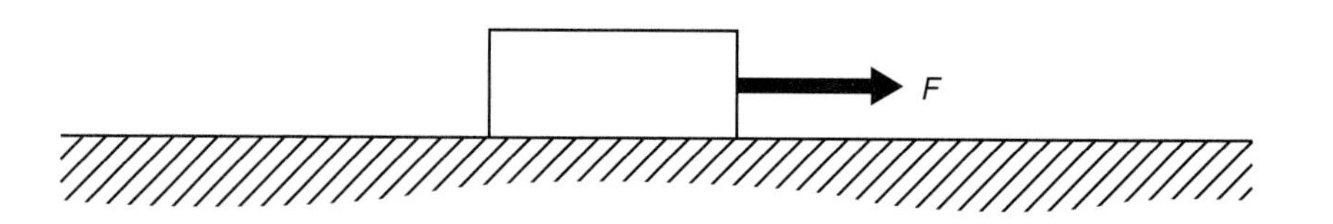

3. Explain how the block can move with a constant velocity even though it has a force applied to it. Is *Newton's first law* violated?

4. Approximately what applied force is needed to keep the block moving with a constant velocity that is twice as large as before? Explain.

5. Suppose that a force F of 3.0 N is applied to the block. Sketch on the axes below the shape of the acceleration–time and velocity–time graphs for the block.

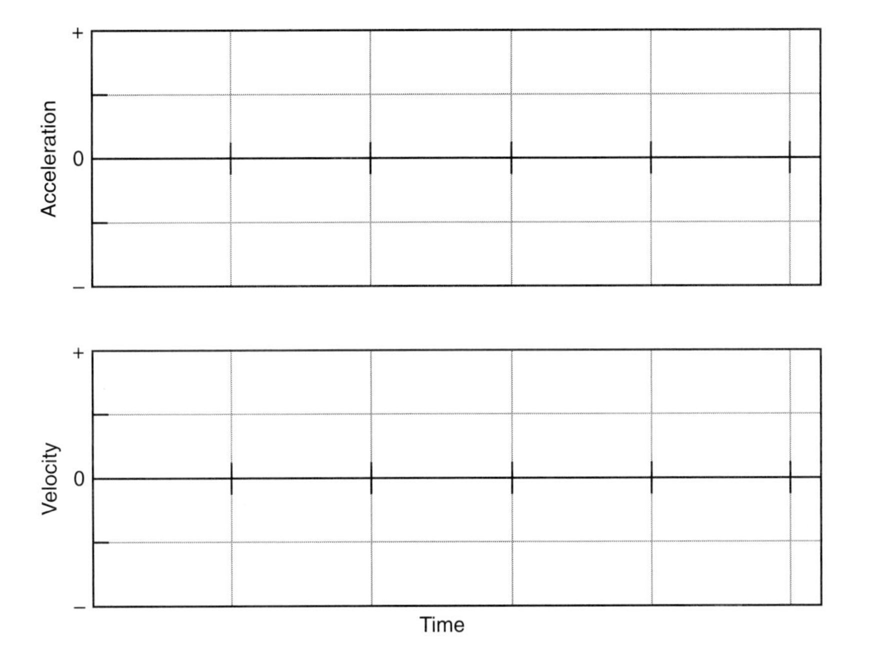

6. In each situation described below, compare the magnitudes of the two forces. Explain your answer in each case.

   a. A 90-kg man and a 60-kg boy each have one hand extended out in front and are pushing on each other. Neither is moving. Compare the force exerted by the man's hand on the boy's hand to that exerted by the boy's hand on the man's.

   b. In (a), the boy begins to slide along the floor. Now compare the same two forces between their hands.

c. A person is leaning against a wall with her hand straight out pushing against the wall. Compare the force exerted by her hand on the wall to that exerted by the wall on her hand. What is the type of force exerted by the wall on her hand called?

d. In (c), is the force exerted by the wall on the person's hand passive or active? Explain.

e. In (c) compare the force exerted by the person's feet on the floor to the force exerted by the floor on her feet. What is the type of force exerted by the floor on her feet called?

f. Is the force the person in (c) exerts on the floor passive or active? Explain.

g. A truck attempts to tow a car. They are connected by a 2-m-long rope. At first the truck doesn't pull hard enough, and the car doesn't move. Compare the force exerted by the truck's bumper on the rope to that exerted by the rope on the truck's bumper. Also compare the force exerted by the rope on the car's bumper to that exerted by the car's bumper on the rope.

h. Finally the truck pulls hard enough so that the car begins to move. Compare the same pairs of forces to each other.

i. An elevator is hanging from a strong cable. The elevator is at rest. Compare the force exerted by the cable on the elevator to that exerted by the elevator on the cable.

j. In (i) compare the tension in the cable to the weight of the elevator.

k. The elevator in (i) begins accelerating upward. Now compare the force exerted by the cable on the elevator to that exerted by the elevator on the cable.

l. In (k) compare the tension in the cable to the weight of the elevator.

m. The elevator in (i) is moving upward at a constant velocity. Now compare the force exerted by the cable on the elevator to that exerted by the elevator on the cable.

n. In (m) compare the tension in the cable to the weight of the elevator.

Name________________________________ Date____________________

Pre-Lab Preparation Sheet for Lab 8: One-Dimensional Collisions

(Due at the beginning of Lab 8)

Directions:
Read over Lab 8 and then answer the following questions about the procedures.

1. What is your Prediction 1-2? Which would be more effective at closing the door—the superball or the clay ball of the same mass?

2. What is the final momentum of the clay ball in Activity 1-3?

3. How do you find the impulse of a force from the force vs. time graph?

4. Why is the collision in Activity 2-2 made with a "springy" wall?

5. How will you measure the impulse in Activity 2-2?

Name________________________ Date____________ Partners________________________

Lab 8: One-Dimensional Collisions

In any system of bodies which act on each other, action and reaction, estimated by momentum gained and lost, balance each other according to the laws of equilibrium.

—Jean de la Rond D'Alembert

OBJECTIVES

- To understand the definition of momentum and its vector nature as it applies to one-dimensional collisions.
- To develop the concept of impulse to explain how forces act over time when an object undergoes a collision.
- To study the interaction forces between objects that undergo collisions.
- To examine the relationship between impulse and momentum experimentally in both *elastic* (bouncy) and *inelastic* (sticky) collisions.

OVERVIEW

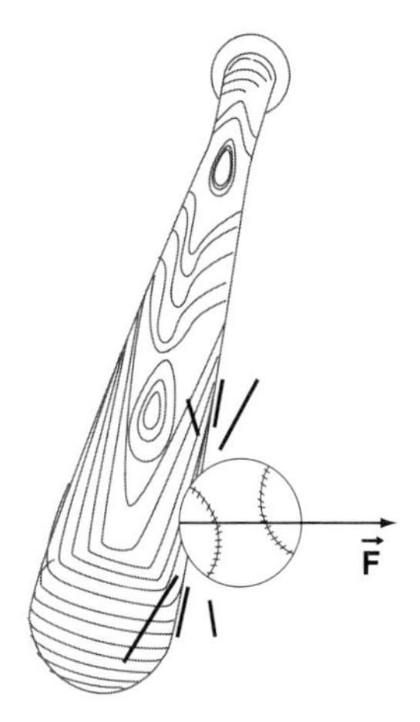

In this lab we explore the forces of interaction between two objects and study the changes in motion that result from these interactions. We are especially interested in studying collisions and explosions in which interactions take place in fractions of a second. Early investigators spent a considerable amount of time trying to observe collisions and explosions, but they encountered difficulties. This is not surprising, since the observation of the details of such phenomena requires the use of instruments—such as high-speed cameras—that were not yet invented.

However, the principles describing the outcomes of collisions were well understood by the late seventeenth century when several leading European scientists, including Isaac Newton, developed the concept of *quantity-of-motion* to describe both *elastic* collisions in which objects bounce off each other and *inelastic* collisions in which objects stick together. These days we use the word *momentum* rather than *quantity-of-motion* to help us understand the nature of collisions and explosions.

We will begin our study of collisions by exploring the relationship between the forces experienced by an object and its momentum change. It can be shown mathematically from Newton's laws and experimentally from our own observations that the change in momentum of an object is equal to a quantity called *impulse.* Impulse takes into account both the magnitude of the applied force at each instant in time and the time interval over which this force acts. The statement of equality between impulse and momentum change is known as the *impulse–momentum theorem.*

INVESTIGATION 1: MOMENTUM AND MOMENTUM CHANGE

In this investigation we are going to develop the concept of momentum to predict the outcome of collisions. But you don't officially know what momentum is because we haven't defined it yet. Let's start by predicting what will happen as a result of a simple one-dimensional collision. This should help you figure out how to define momentum to enable you to describe collisions in mathematical terms.

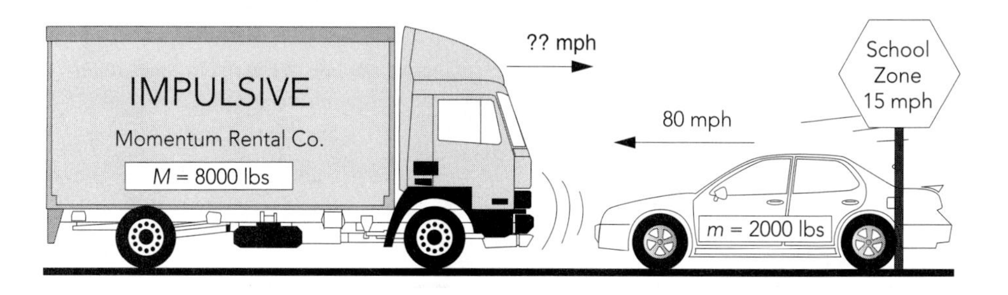

It's early fall and you are driving along a two-lane highway in a rented moving van. It's full of all of your possessions, so you and the loaded truck weigh 8000 lb. You have just slowed down to 15 mph because you're in a school zone. It's a good thing you thought to do that because a group of first graders are just starting to cross the road. Just as you pass the children, you see a 2000-lb sports car in the other lane heading straight for the children at about 80 mph.

A desperate thought crosses your mind. You just have time to swing into the other lane and speed up a bit before making a head-on collision with the sports car. *You want your truck and the sports car to crumple into a heap that sticks together and doesn't move.* Can you save the children or is this just a suicidal act?

To simulate this situation you can observe two carts of different mass set up to stick together in trial collisions. You will need

- low-friction cart
- low-friction cart of four times the mass
- Velcro or clay to make the carts stick together
- level track or other smooth level surface

Activity 1-1: Can You Stop the Car?

Prediction 1-1: How fast would you have to be going to completely stop the sports car? Explain the reasons for your prediction.

1. Try some head-on collisions with the carts of different mass to simulate the event on a small scale. Be sure that the carts stick together after the collision.

2. Observe *qualitatively* what combinations of velocities cause the two carts to be at rest after the collision.

Question 1-1: What happens when the less massive cart is moving much faster than the more massive cart? Much slower? At an intermediate speed?

Question 1-2: Based on your prediction and your observations, what mathematical definition might you use to describe the momentum you would need to stop an oncoming vehicle traveling with a known mass and velocity? Should it depend on the mass, the velocity, or both? Explain your choice.

Just to double-check your reasoning, you should have come to the conclusion that momentum is defined by the vector equation

$$\vec{p} \equiv m\vec{v}$$

where the symbol $\equiv$ is used to designate "defined as."

Originally, Newton did not use the concept of acceleration or velocity in his laws. Instead, he used the term "motion," which he defined as the product of mass and velocity (the quantity we now call momentum). Let's examine a translation from Latin of Newton's first two laws with some parenthetical changes for clarity.

NEWTON'S FIRST TWO LAWS OF MOTION*, †

1. Every body continues in its state of rest, or of uniform motion in a right line, unless it is compelled to change that state by forces impressed on it.
2. The (rate of) change of motion is proportional to the motive force impressed: and is made in the direction of the right line in which that force is impressed.

The more familiar contemporary statement of the second law is that the net force on an object can be calculated as the product of its mass and its acceleration where the direction of the force and of the resulting acceleration are the same. Newton's statement of the second law and the more modern statement are mathematically equivalent.

Now let's test your intuition about momentum and forces. You are sleeping in your sister's room while she is away at college. Your house is on fire and smoke is pouring into the partially open bedroom door. To keep the smoke from coming in, you must close the door. The room is so messy that you cannot get to the door. The only way to close the door is to throw either a blob of clay or a superball at the door—there isn't time to throw both.

*I. Newton, *Principia Mathematica,* Florian Cajori, ed. (Berkeley: University of California Press, 1934), p. 13.
†L.W. Taylor, *Physics: The Pioneer Science,* Vol. 1 (New York: Dover, 1959), pp. 129–131.

Activity 1-2: Which Packs the Bigger Wallop—A Clay Blob or a Superball?

Prediction 1-2: Assuming the clay blob and the superball have the same mass, and that you throw them with the same velocity, which would you throw to close the door—the clay blob, which will stick to the door, or the superball, which will bounce back at almost the same speed as it had before it collided with the door? Give reasons for your choice using any notions you already have or any new concepts developed in physics, such as force, energy, momentum, or Newton's laws. If you think that there is no difference, justify your answer. Remember, your life depends on it!

You can test your prediction by dropping a force probe with a bouncy rubber stopper on its end (instead of a superball), and then dropping the force probe from the same height with a clay ball of approximately the same mass on its end. We can associate the maximum force read by the force probe with the maximum force a thrown ball can exert on the door. We will later investigate how the force is related to the change in momentum of the ball in each case. To do these observations you'll need the following equipment:

- computer-based laboratory system
- *RealTime Physics Mechanics* experiment configuration files
- force probe
- rubber stopper with a hole in it
- round blob of clay of the same mass as the stopper
- additional clay
- meter stick

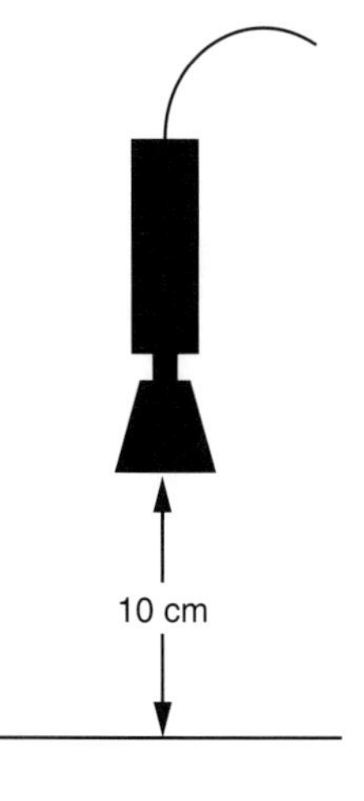

Let's compare the maximum forces imparted to the force probe for the two types of collisions.

1. Open the experiment file called **Clay vs. Superball (L08A1-2).** This will set up the computer to collect and graph force data at 4000 points per second in triggered mode with a push as positive, on the axes that follow.
2. Mount the rubber stopper on the end of the force probe. (If you are using a Hall effect force probe, you may need to adjust the spacing and **check the sensitivity.**)
3. **Zero** the force probe while holding it in a vertical position with the stopper pointing down. Begin with the rubber stopper 10 cm above the table. (Record the height in Table 1-2.)

Table 1-2

	Mass (kg) (including force probe)	Height (m)	Maximum force (arbitrary units)
Stopper			
Small clay ball			
More massive clay ball			
Small clay ball, larger height			

4. **Begin graphing,** and then drop the force probe. Repeat this several times, **zeroing** the force probe before each measurement. *Be sure that the force probe falls vertically downword and doesn't tip to one side.*

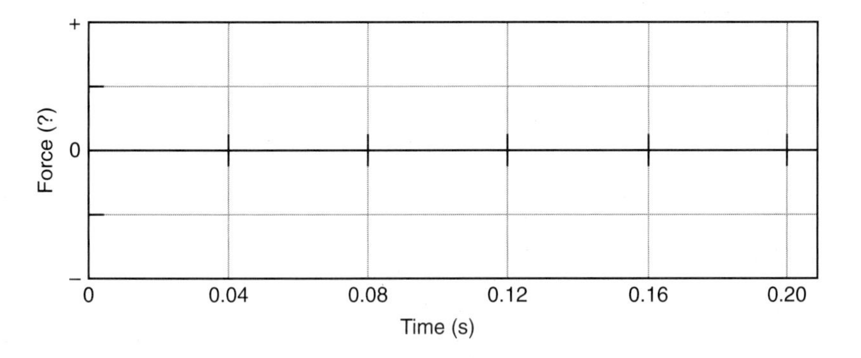

5. Move the data from your last good run so that the graphs are **persistently displayed on the screen** for later comparison.
6. Use the **analysis feature** of the software to find the maximum force applied to the force probe, and record it in Table 1-2.
7. Now replace the stopper with a ball of clay of about the same mass. Be sure to **zero** the force probe with the clay pointing vertically downward before beginning to graph.
8. Drop the force probe from the same height, find the maximum force and record it in the table. Also record the masses of the stopper plus force probe and clay plus force probe, and the height of the drop.
9. Sketch both graphs on the previous axes, or **print** the graphs and affix them over the axes.

Question 1-3: Did your observations agree with your prediction? Which resulted in a bigger maximum force—the stopper or clay?

Question 1-4: Based on your observations, which should you throw at the door—the superball or the clay? Explain.

Prediction 1-3: What will happen to the maximum force if you increase the mass of the ball but allow it to collide with the same velocity (drop it from the same height)?

Prediction 1-4: What will happen to the maximum force if the velocity just before impact is increased by dropping the ball from a greater height?

10. Test your Prediction 1-3 by using a more massive clay ball dropped from the same height, so that the mass of the clay ball and force probe is about doubled. Measure the maximum force and record it in Table 1-2.
11. Test your Prediction 1-4 by dropping the smaller clay ball from about twice the height. Record the height and maximum force in Table 1-2.

Question 1-5: Did your observations agree with your predictions? What factors seem to determine the force exerted on the force probe?

It would be nice to be able to use Newton's formulation of the *second law* of motion to find collision forces, but it is difficult to measure the rate of change of momentum during a rapid collision without special instruments. However, measuring the momenta of objects just before and just after a collision is not usually too difficult. This led scientists in the seventeenth and eighteenth centuries to concentrate on the overall changes in momentum that resulted from collisions. They then tried to relate changes in momentum to the forces experienced by an object during a collision.

In the next activity you are going to explore the mathematics of calculating momentum changes for the two types of collisions—the *elastic* collision, where the ball bounces off the door, and the *inelastic* collision, where the ball sticks to the door.

Activity 1-3: Momentum Changes

Prediction 1-5: Which object undergoes the greater *momentum change* during the collision with a door—the clay blob or the superball? Explain your reasoning carefully.

Comment: Recall that momentum is defined as a *vector* quantity; i.e., it has both *magnitude* and *direction*. Mathematically, momentum change is given by the equation

$$\Delta\vec{p} = \vec{p}_{\mathrm{f}} - \vec{p}_{\mathrm{i}}$$

where $\vec{p}_{\mathrm{i}}$ is the initial momentum of the object just before a collision and $\vec{p}_{\mathrm{i}}$ is its final momentum just after. Remember, in one dimension, the *direction* of a vector is indicated by its *sign*.

Check your prediction with some calculations of the momentum changes for both collisions that you carried out. This is a good review of the properties of one-dimensional vectors. Carry out the following calculations for the *original height* and *original mass* of both the stopper and clay ball.

1. Calculate the initial momentum of the clay ball plus force probe just before it hits the table. (**Hint:** You will need to recall from kinematics with constant acceleration that $v = (2a_g h)^{1/2}$, where $a_g = 9.8\ \text{m/s}^2$ is the gravitational acceleration, and h is the distance the ball falls before hitting the table.) Take the positive y axis as *upward*. Show your calculation.

$$p_i =$$

2. What is the final momentum of the clay ball and force probe after it collides with the table? Explain.

$$p_f =$$

3. What is the change in momentum of the clay ball and force probe? Be careful of the sign.

$$\text{Clay ball:} \qquad \Delta p =$$

4. Now calculate the change in momentum of the stopper and force probe from just before it hits the table until just after it bounces up from the table. Assume that the stopper bounces in such a way that the *magnitude* of its velocity doesn't change. Show your calculation below. Be very careful of signs!

$$\text{Stopper:} \qquad \Delta p =$$

Question 1-6: Compare your calculated changes in momentum to your predictions. Do they agree? Which ball had the larger change in momentum?

Question 1-7: How does change in momentum seem to be related to the maximum force applied to the ball?

You have observed that the ball that bounces off the door will exert a larger maximum force on the door than the clay blob, which sticks to it. The ball that bounces off the door has the larger change in momentum because it reverses direction, while the clay blob merely comes to rest stuck to the door.

However, just looking at the *maximum* force exerted on the ball does not tell the whole story. You can see this from a simple experiment tossing raw eggs. We will do it as a thought experiment to avoid the mess! Suppose somebody tosses you a raw egg and you catch it. (In physics jargon, one would say in a very official tone of voice, "The egg and the hand have undergone an *inelastic collision.*") What is the relationship between the force you have to exert on the egg to stop it, the time it takes you to stop it, and the momentum change that the egg experiences? You ought to have some intuition about this matter. In more ordinary language, would you want to catch the egg slowly (by relaxing your hands and pulling them back) or quickly (by holding your hands rigidly)?

Activity 1-4: Momentum Change and Force on an Egg

Question 1-8: If you catch an egg of mass m that is heading toward your hand at speed v, what magnitude momentum *change* does it undergo? (**Hint:** The egg is at rest after you catch it.)

Question 1-9: Does the total momentum change differ if you catch the egg more slowly or is it the same?

Question 1-10: Suppose the time you take to bring the egg to a stop is Δt. Would you rather catch the egg in such a way that Δt is small or large? Why?

Question 1-11: What do you suspect might happen to the average force you exert on the egg while catching it when Δt is small?

In bringing an egg to rest, the change in momentum is the same whether you use a large force during a short time interval or a small force during a long time interval. *Of course, which one you choose makes a lot of difference in whether the egg breaks or not!*

A quantity called *impulse* may have been defined for you in lecture and/or in your textbook. It combines the applied force and the time interval over which it acts. *In one dimension,* for a *constant* force F acting over a time interval Δt, as shown in the graph below, the impulse J is

$$J = F\Delta t$$

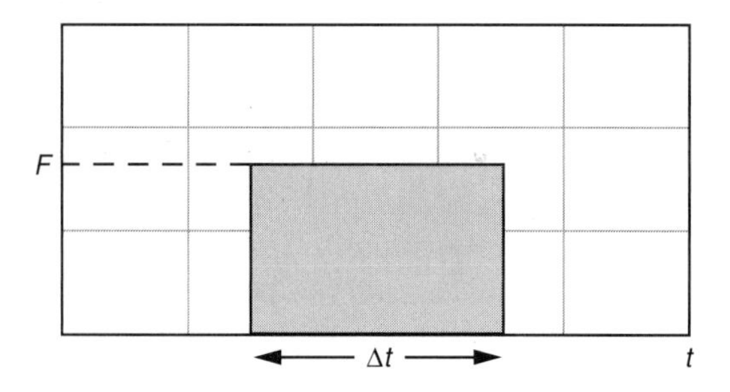

As you can see, a large force acting over a short time and a small force acting over a long time can have the same impulse.

Note that $F\Delta t$ is the area of the rectangle, i.e., the area under the force vs. time curve. If the applied force is not constant, then the impulse can still be calculated as the *area under the force vs. time graph.*

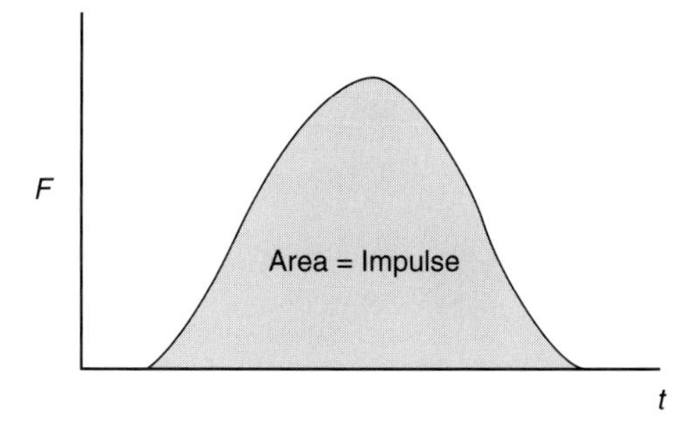

It is the impulse that equals the change in momentum. *In one dimension,*

$$J = \Delta p$$

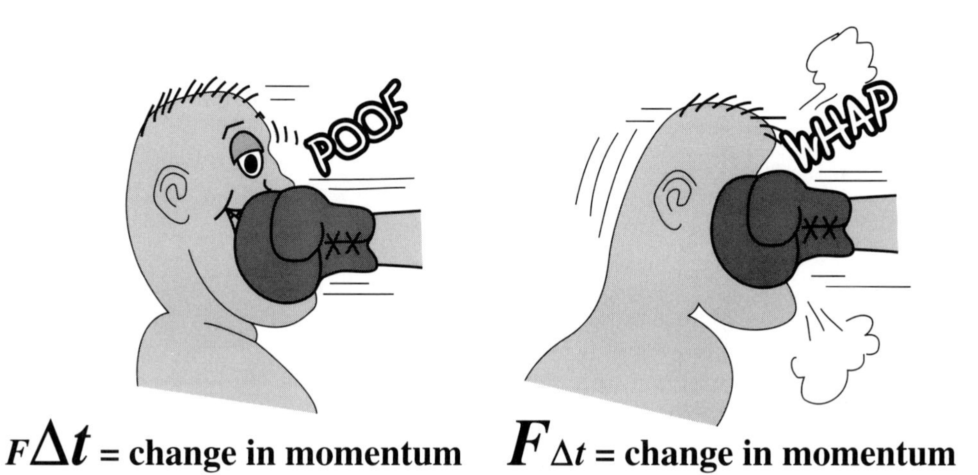

INVESTIGATION 2: IMPULSE, MOMENTUM, AND COLLISIONS

Let's first see qualitatively what an impulse curve might look like in a real collision in which the forces change over time during the collision. To explore this idea you will need

- low-friction cart with a spring plunger

Activity 2-1: Observing Collision Forces That Change With Time

Collide the cart with the wall several times and observe what happens to the spring plunger.

Question 2-1: If friction is negligible, what is the net force exerted on the cart *just before* it starts to collide?

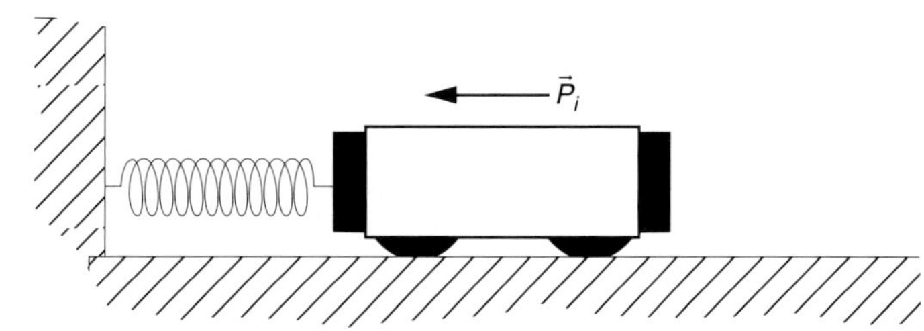

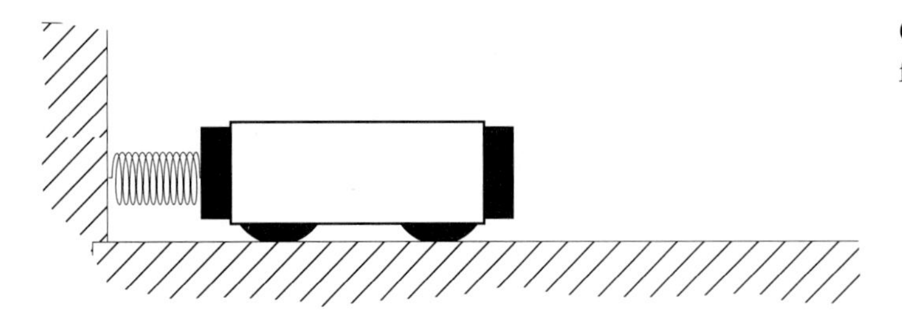

Question 2-2: When is the magnitude of the force on the cart maximum?

Question 2-3: Roughly how long does the collision process take? Half a second? Less time? Several seconds?

Prediction 2-1: Remembering what you observed, attempt a rough sketch of the shape of the force the wall exerts on the cart $F_{W \to C}$ as a function of time during the collision.

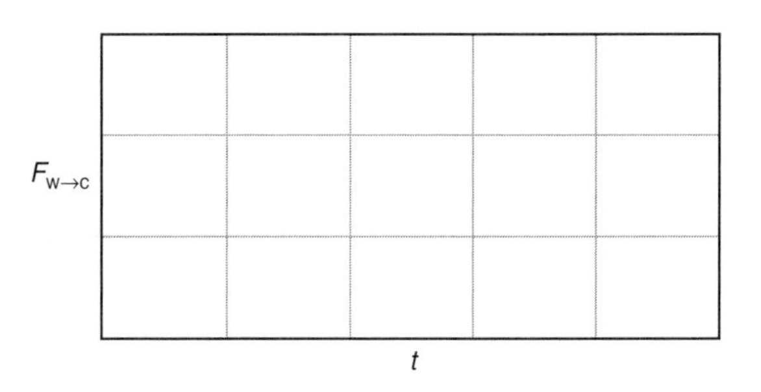

During the collision the force is not constant. To measure the impulse and compare it to the change in momentum of the cart, you must (1) plot a force–time graph and find the area under it, and (2) measure the velocity of the cart before and after the collision with the wall. Fortunately, the force probe, motion detector, and motion software will allow you to do this.

The force probe will be mounted on the cart to measure the *force applied to* the cart. Instead of using a plunger mounted on the cart, you can make the collision nearly *elastic,* and stretch it out over time by colliding the cart into a "springy wall." You can collide the cart into a spring mounted on the wall or into a springy material like foam rubber or a large springy ball.

You will need:

- low-friction cart
- smooth level ramp or other level surface
- computer-based laboratory system
- *RealTime Physics Mechanics* experiment configuration files
- motion detector
- force probe with rubber stopper replacing the hook
- springy wall (spring, foam rubber, or springy ball)

Activity 2-2: Examining the Impulse–Momentum Theorem in a *Nearly Elastic* Collision

In a perfectly elastic collision between a cart and a wall, the cart would recoil with exactly the same magnitude of momentum that it had before the collision. Because your cart's spring bumper is not perfect, you can only produce a *nearly* elastic collision.

1. Fasten the force probe securely to the cart so that the rubber stopper extends beyond the front of the cart.

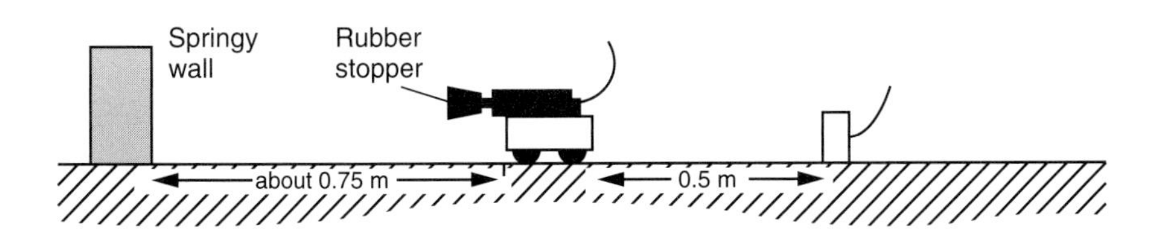

2. Set up the motion detector as shown. *Be sure that the ramp is level.*
3. Measure the mass of the cart and force probe combination.

   Mass of cart plus force probe:______kg
4. Open the experiment file called **Impulse and Momentum (L08A2-2)** to display the axes that follow. This experiment has been set up to record force and motion data at 50 data points per second. Because the positive direction is toward the right, the software has been set up to record a *push* on the force probe as a *positive* force, and velocity *toward the motion detector as positive.*
5. **Calibrate** the force probe *for a push* of 9.8 N by holding the cart with the rubber stopper pointing up and balancing a 1.0-kg mass (9.8 N weight) on the stopper, or **load the calibration.** (If you are using a Hall effect force probe, you may need to adjust the spacing and **check the sensitivity.**)

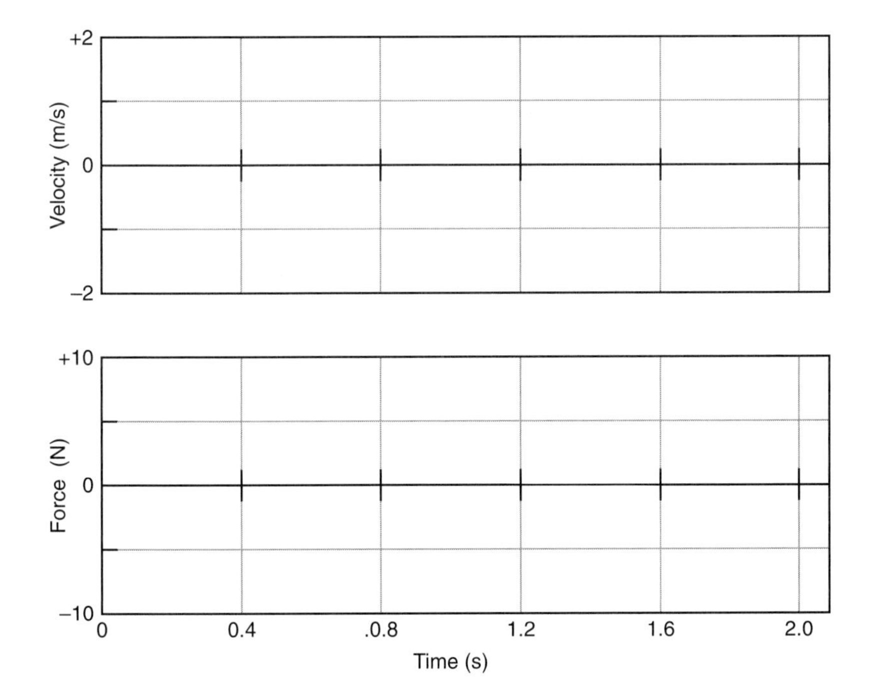

6. Be sure that the wire from the force probe is taped out of the way, so that it won't be seen by the motion detector.

7. Practice pushing the cart toward the wall and watching it bounce off. Find a way to push without putting your hand between the motion detector and the cart.

8. When you are ready, **zero** the force probe and then **begin graphing.** As soon as you hear the clicking of the motion detector, give the cart a push toward the wall, release it, and let it collide.

   Repeat until you get a good set of graphs, i.e., a set in which the motion detector saw the relatively constant velocities of the cart as it moved toward the wall and as it moved away, and also the maximum force was no more than 10 N. (With too large a force, the force probe may read inaccurately.)

Question 2-4: Does the shape of the force–time graph agree with your Prediction 2-1? Explain.

9. Use the **analysis and statistics features** in the software to measure the average velocity of the cart as it approached the wall, and the average velocity as it moved away from the wall.

   Don't forget to include a sign. Positive velocity should be *away from* the wall.

   Average velocity toward the wall:______m/s

   Average velocity away from the wall:______m/s

10. Calculate the change in momentum of the cart. Show your calculations.

$$\Delta p = \_\_\_\_\_\_ \text{kg m/s}$$

11. Use the **integration routine** in the software to find the area under the force–time graph—the impulse. (The area under a curve is the same as the *integral* of force vs. time.)

$$J = \_\_\_\_\_\_ \text{N} \cdot \text{s}$$

12. Sketch your graphs on the previous axes, or **print** them and affix them over the axes.

Question 2-5: Did the calculated change in momentum of the cart equal the measured impulse applied to it by the wall during the nearly elastic collision? Explain.

What would the impulse be if the initial momentum of the cart were larger? What if the collision were inelastic rather than elastic, i.e., what if the cart stuck to the wall after the collision? If you have more time, do the following Extensions to find answers to these questions.

Extension 2-3: A Larger Momentum Change

Suppose a more massive cart collided nearly elastically with the wall. What would the impulse be? You can add mass to the cart and find out.

Prediction E2-2: If the cart had twice the mass and collided with the wall elastically moving at the same velocity as in Activity 2-2, how large do you think the impulse would be? The same as before? Larger? Smaller? Why?

Test your prediction. Add masses to your cart to make the total mass twice as large.

New mass of cart:______kg

You can use the same experiment file, **L08A2-2 (Impulse and Momentum),** as in Activity 2-2.

Zero the force probe, and then collide the cart with the springy wall again. Try several times until you get the initial velocity about the same as in Activity 2-2.

Find the average velocities, as in Activity 2-2, and calculate the change in momentum.

Average velocity toward the wall: ______m/s

Average velocity away from the wall: ______m/s

$\Delta p =$ ______kg m/s

Find the impulse as in Activity 2-2. **Print** your graphs and affix them below, or sketch them on the axes.

$J =$ ______N · s

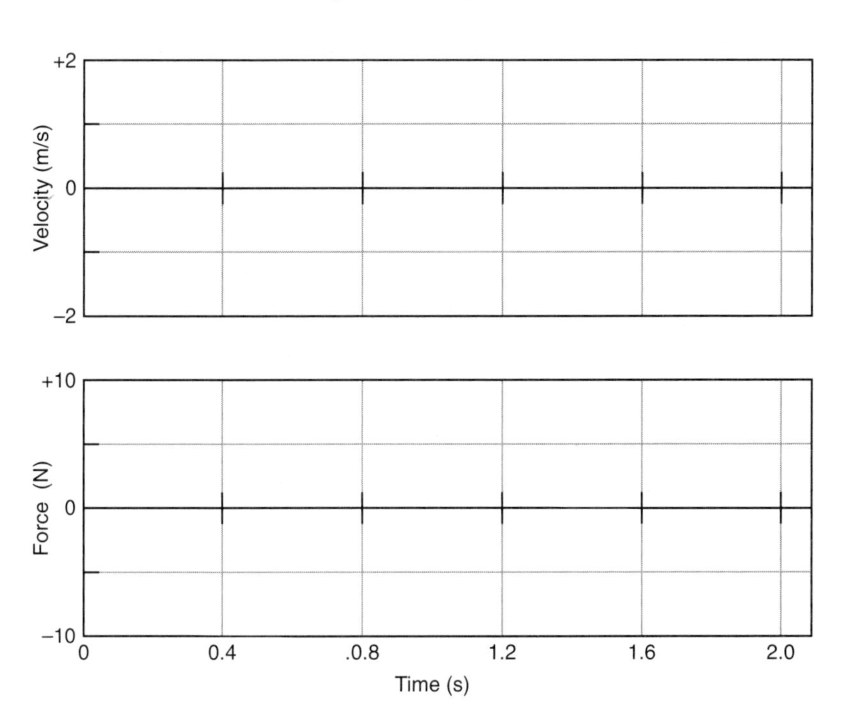

Question E2-6: Did the impulse agree with your prediction? Explain.

Question E2-7: Were the impulse and change in momentum equal to each other? Explain why you think the results came out the way they did.

Extension 2-4: Impulse–Momentum Theorem in an Inelastic Collision

It is also possible to examine the impulse–momentum theorem in a collision where the cart sticks to the wall and comes to rest after the collision. This can be done by replacing the bouncy wall with some clay, and attaching a nail to the end of the force probe.

In addition to the equipment you have used so far, you will need

- long nail
- large blob of clay

Leave the extra mass on your cart so that its mass is the same as in Extension 2-3. Attach the nail to the stopper. (If you are using a Hall effect force probe, *be careful not to rotate the head of the force probe, since this will change the calibration.*)

Remove the springy material and attach a blob of clay to the wall at the height of the nail. The rest of the setup is as in Activity 2-2.

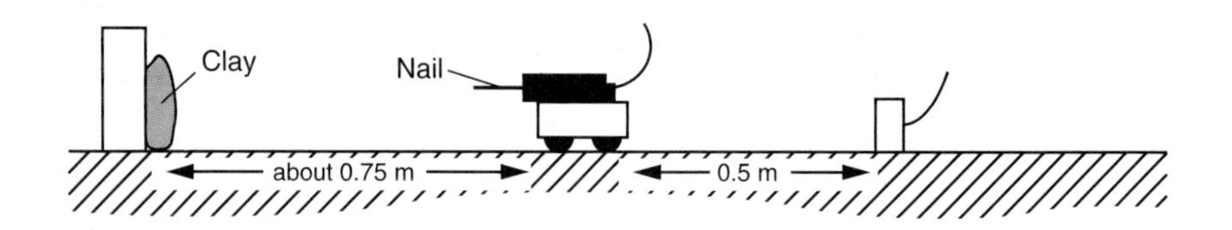

Prediction E2-3: Now when the cart hits the wall, it will come to rest stuck to the clay. What do you predict about the impulse? Will it be the same, larger, or smaller than in the nearly elastic collision? What do you predict now about the impulse and momentum? Will they equal each other, or will one be larger than the other?

You can use the same experiment file, **Impulse and Momentum (L08A2-2),** as in Activity 2-2.

Zero the force probe, and then collide the cart with the clay. Try several times until you get the initial velocity about the same as in Activity 2-2.

Find the average velocity, as in Activity 2-2, and calculate the change in momentum.

Average velocity toward the wall: ______m/s

$\Delta p =$ ______kg m/s

Find the impulse as in Activity 2-2. **Print** your graphs and affix them over the axes that follow, or sketch them on the axes.

$J =$ ______ $\mathrm{N \cdot s}$

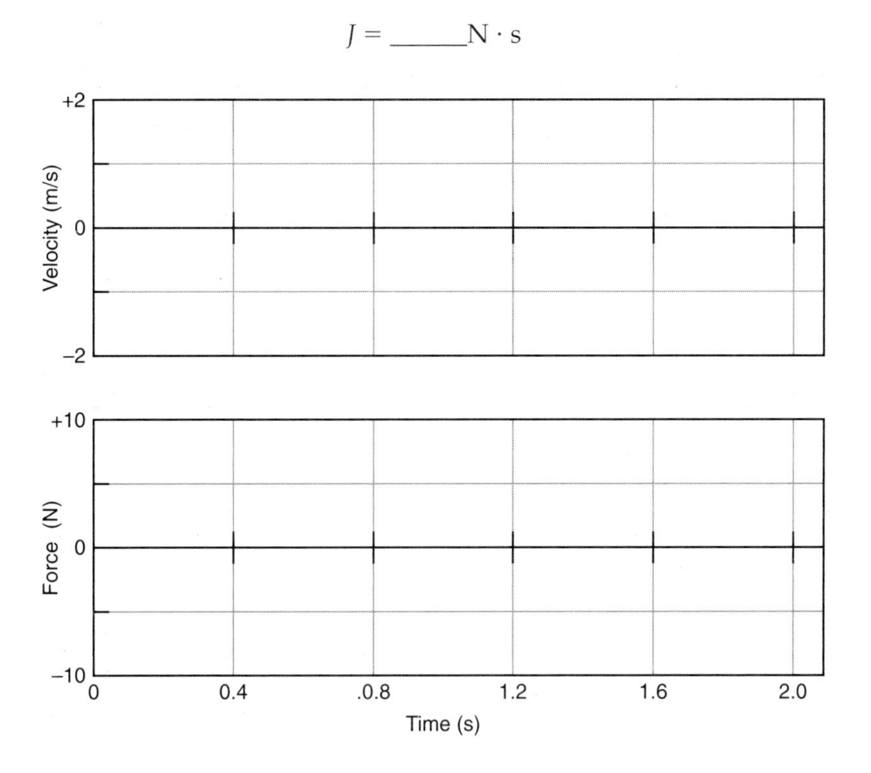

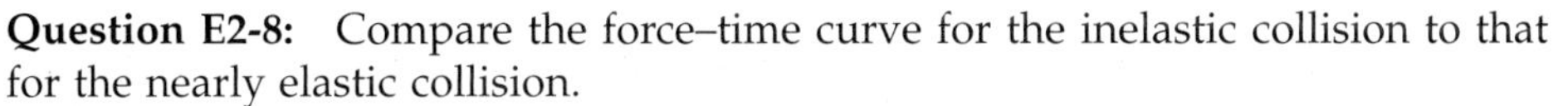

Question E2-8: Compare the force–time curve for the inelastic collision to that for the nearly elastic collision.

Question E2-9: Were the impulse and change in momentum equal to each other for the inelastic collision? Explain why you think the results came out the way they did.

Question E2-10: Do you think that the momentum change is equal to the impulse for all collisions? Justify your answer.

Name____________________ Date__________ Partners____________________

Homework for Lab 8: One-Dimensional Collisions

1. Find the impulse of the force shown on the force–time graph below. Explain how you found your answer.

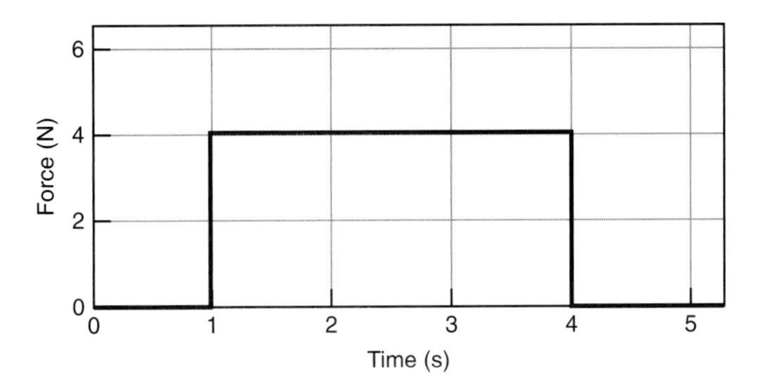

2. An object of mass 2.5 kg is moving in the negative x direction at a velocity of 2.0 m/s. It experiences the force shown above for 3 S. What is the final velocity after the object has experienced the impulse. Show your calculations.

3. A ball of mass 1.5 kg is thrown upward. It leaves the thrower's hand with a velocity of 10 m/s. The following questions refer to the motion *after the ball leaves the thrower's hand.* Assume that the upward direction is positive. Show all calculations.

   a. How long does it take for the ball to return to the thrower's hand?

   b. What is the final velocity of the ball just before it reaches the hand?

   c. What is the change in momentum of the ball?

d. What is the impulse calculated from the change in momentum?

e. What is the average force acting on the ball?

4. After the ball in Question 3 hits the thrower's hand, it comes to rest in a time of 0.25 s.

a. What is the net impulse exerted on the ball?

b. What is the average force exerted by the hand on the ball? (**Hint:** Don't forget the gravitational force.)

5. A superball of mass 0.05 kg is dropped from a height of 10 cm above a table top. It bounces off the table and rises to the same height.

Sketch the shape of the force exerted on the ball by the table as a function of time on the axes on the right.

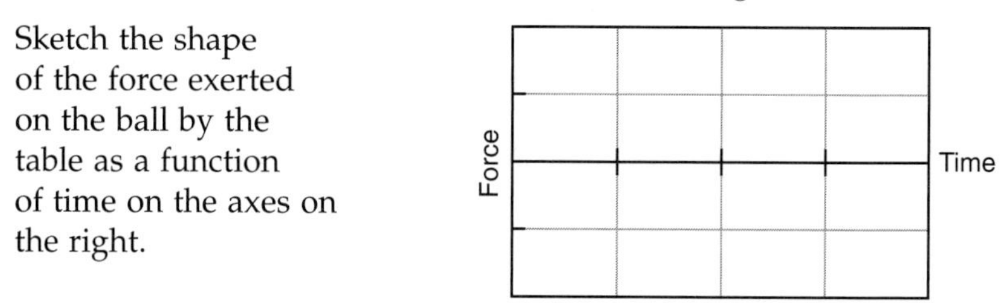

6. If the superball was in contact with the table for 30 ms, calculate the average force exerted *on the ball by the table.* Show your work. (**Hint:** First calculate the momentum before and after hitting the table. Don't forget the gravitational force.)

Average force: ______

7. The superball is now replaced by a clay ball the same size and mass. The ball is dropped from the same height and it sticks to the table.

Sketch the shape of the force exerted on the ball by the table as a function of time on the axes on the right.

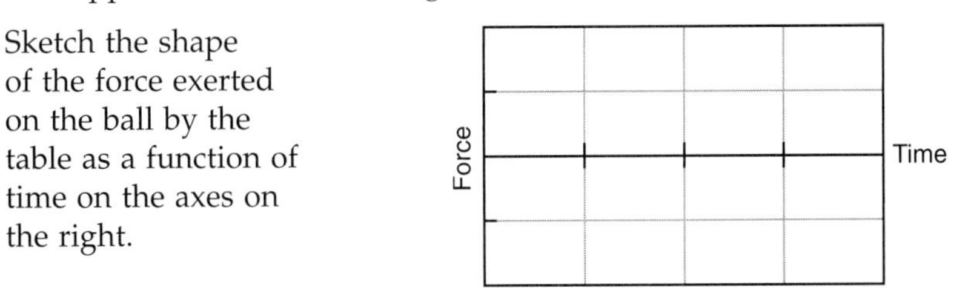

8. Calculate the impulse exerted on the clay ball and compare it to that with the superball. Which is larger or are they both the same? Explain.

9. If the collision of the clay ball with the table takes the same 30 ms as the collision of the superball, calculate the average force exerted *by the table on the clay ball* and compare it to that exerted on the superball. Which is larger or are they the same? Explain.

10. Suppose that the clay has twice the mass and is dropped from the same height. Compare the impulse exerted on the ball by the table to that with the smaller clay ball. Explain your answer.

11. Suppose that the original clay ball is dropped from twice the height. Compare the impulse exerted on the ball by the table to that for the smaller height. Explain your answer.

Name________________________________ Date____________________

Pre-Lab Preparation Sheet for Lab 9: Newton's Third Law And Conservation Of Momentum

(Due at the beginning of Lab 9)

Directions:
Read over Lab 9 and then answer the following questions about the procedures.

1. Write down the definition of momentum.

2. What is your Prediction 1-2? Does one car exert a larger force on the other or are both forces the same size?

3. In Activity 1-1, why is the sign of force probe A reversed?

4. What is your Prediction 1-7 when the truck is accelerating? Does either the car or the truck exert a larger force on the other or are the forces the same size?

5. What makes the collision in Activity 2-1 "inelastic"?

REALTIME PHYSICS: MECHANICS

Name______________________ Date__________ Partners______________________

Lab 9: Newton's Third Law and Conservation of Momentum

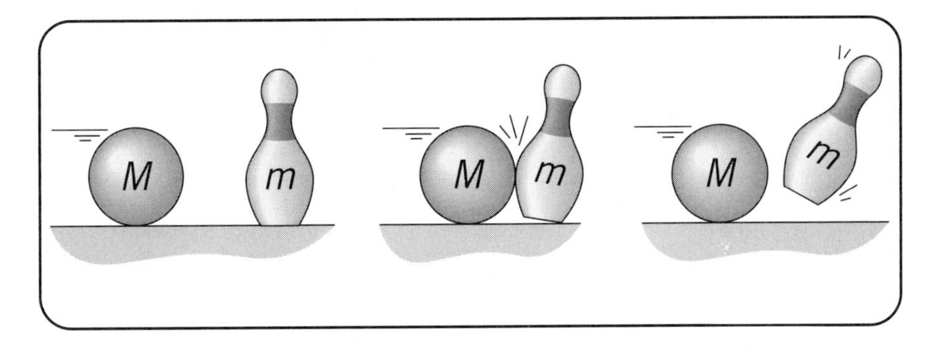

To every action there is always opposed an equal reaction, or the mutual actions of two bodies upon each other are always equal, and directed to contrary parts.

If you press a stone with your finger, the finger is also pressed by the stone. If a horse draws a stone tied to a rope, the horse (if I may say so) will be equally drawn back towards the stone . . . .

—Isaac Newton

OBJECTIVES

- To study the forces between objects that interact with each other, especially in collisions.
- To examine the consequences of *Newton's third law* as applied to interaction forces between objects.
- To formulate the *law of conservation of momentum* as a theoretical consequence of *Newton's third law* and the impulse–momentum law.
- To explore conservation of momentum in one-dimensional collisions.

OVERVIEW

In Lab 8, you looked at the definition of momentum and examined the momentum changes of objects undergoing collisions. We focused our attention on the momentum change that an object undergoes when it experiences a force that is extended over time (even if that time is very short!). You also looked at the forces acting on a single object during a collision and examined the impulse–momentum law, which compares the change in the object's momentum to the impulse it experiences.

Since interactions like collisions and explosions never involve just one object, we turn our attention to the mutual forces of interaction between two or more objects. This will lead us to a very general law known as *Newton's third law,* which relates the forces of interaction exerted by two objects on each other. Then, you will examine the consequences of this law and the impulse–momentum law, which you examined in the last lab, when they are applied to collisions between objects. In doing so, you will arrive at one of the most important laws of interactions between objects, the *conservation of momentum law.*

As usual you will be asked to make some predictions about interaction forces and then be given the opportunity to test these predictions.

INVESTIGATION 1: FORCES BETWEEN INTERACTING OBJECTS

There are many situations where objects interact with each other, for example, during collisions. In this investigation we want to compare the forces exerted by the objects on each other. In a collision, both objects might have the same mass and be moving at the same speed, or one object might be much more massive, or they might be moving at very different speeds. What factors might determine the forces the objects exert on each other? Is there some general law that relates these forces?

Activity 1-1: Collision Interaction Forces

What can we say about the forces two objects exert on each other during a collision?

Prediction 1-1: Suppose two objects have the same mass and are moving toward each other at the same speed so that $m_A = m_B$ and $\vec{v}_A = -\vec{v}_B$ (same speed, opposite direction).

$$m_A = m_B \quad \text{and} \quad \vec{v}_A = -\vec{v}_B$$

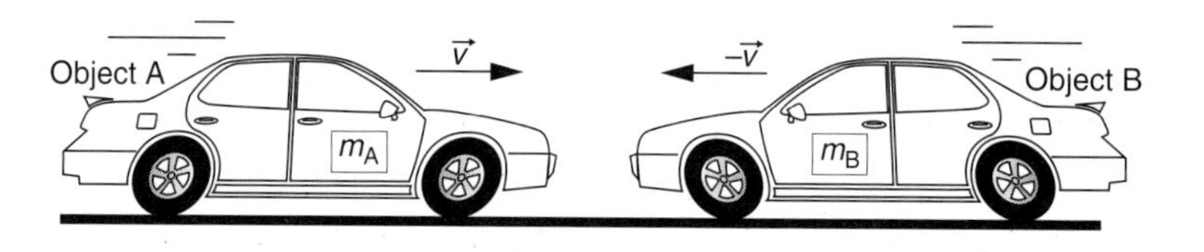

Predict the relative magnitudes of the forces between object A and object B during the collision. Place a check next to your prediction.

______Object A exerts a larger force on object B.

______The objects exert the same size force on each other.

______Object B exerts a larger force on object A.

Prediction 1-2: Suppose the masses of two objects are the same and that object A is moving toward object B, but object B is at rest.

$$m_A = m_B \quad \text{and} \quad \vec{v}_A \neq 0, \vec{v}_B = 0$$

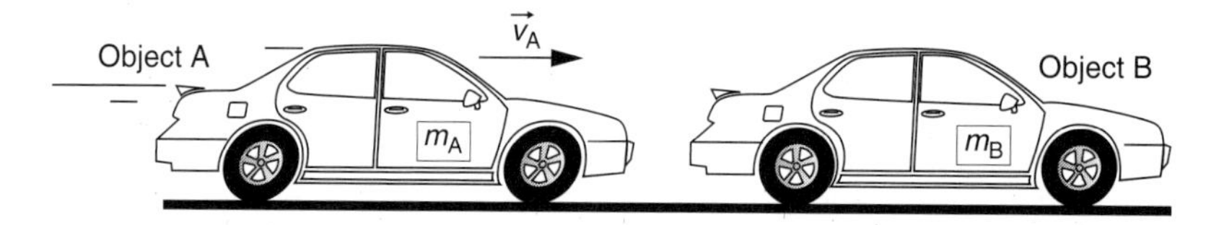

Predict the relative magnitudes of the forces between object A and object B during the collision.

______Object A exerts a larger force on object B.

______The objects exert the same size force on each other.

______Object B exerts a larger force on object A.

Prediction 1-3: Suppose the mass of object A is greater than that of object B and that it is moving toward object B, which is at rest.

$$m_A > m_B \quad \text{and} \quad \vec{v}_A \neq 0, \vec{v}_B = 0$$

Predict the relative magnitudes of the forces between object A and object B during the collision.

______Object A exerts a larger force on object B.

______The objects exert the same size force on each other.

______Object B exerts a larger force on object A.

Provide a summary of your predictions. What are the circumstances under which you predict that one object will exert a greater force on the other object?

To test the predictions you made you can study *gentle* collisions between two force probes attached to carts. You can add masses to one of the carts so it has significantly more mass than the other. If a compression spring is available you can also study an "explosion" between the two carts by compressing the spring between the force probes on each cart and letting it go. To make these observations of interactions you will need the following equipment:

- computer-based laboratory system
- *RealTime Physics Mechanics* experiment configuration files
- two force probes with rubber stoppers replacing the hooks
- two 1-kg masses to calibrate the force probes
- two low-friction carts
- masses to place on one of the carts to double and triple its mass

- smooth level ramp or other level surface
- small spring that can be compressed (optional)

1. Set up the apparatus as shown in the following diagram.

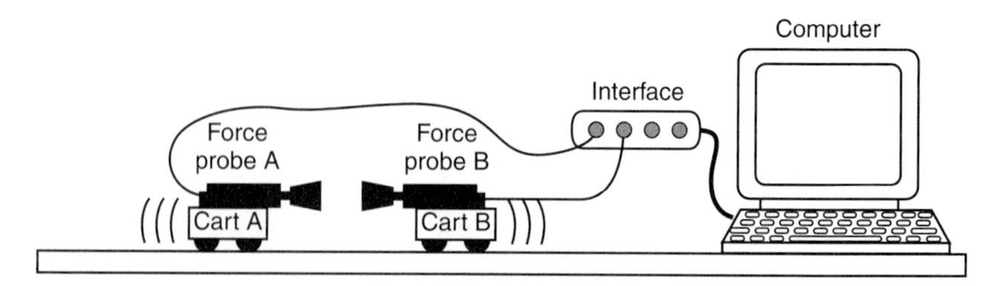

The force probes should be securely fastened to the carts.

The hooks should be removed from the force probes and replaced by rubber stoppers, which should be *carefully aligned* so that they will collide head-on with each other.

If the carts have friction pads, these should be raised so that they don't rub on the ramp.

2. Open the experiment file called **Collisions (L09A1-1)** to display the axes shown below. The software will be set up to measure the forces applied to each probe with a very fast data collection rate of 4000 points per second. (This allows you to see all of the details of the collision which takes place in a very short time interval.) The software will also be set up to be **triggered,** so that data collection will not start until the carts actually collide.

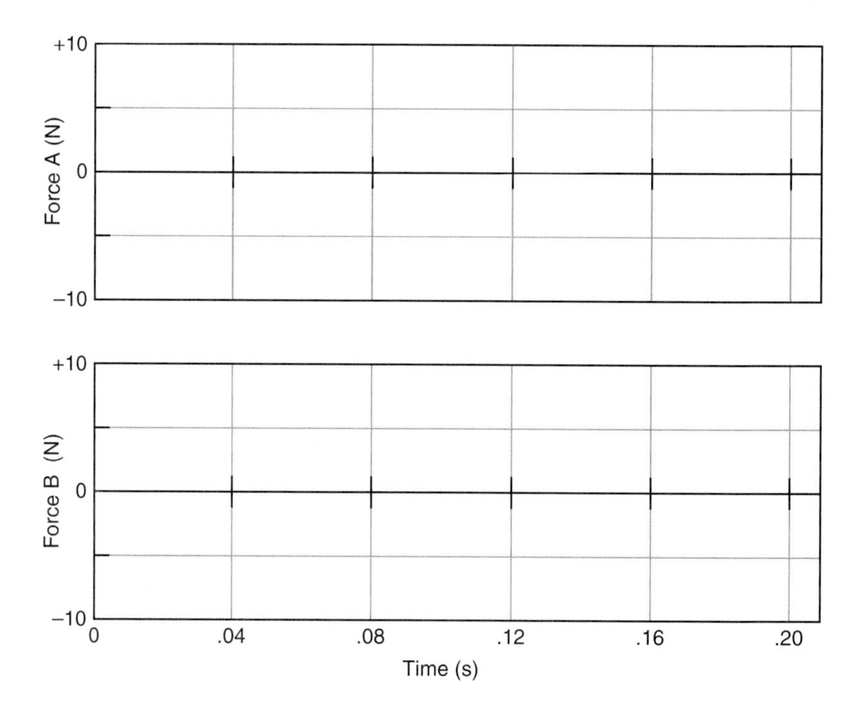

3. **Calibrate** both force probes *for pushes* with 1.0-kg (9.8-N) masses balanced on each stopper, or **load the calibrations.** You may find it easier to calibrate one force probe at a time. (If you are using a Hall effect force probe, you may need to adjust the spacing and **check the sensitivity.**)

4. **Reverse the sign** of force probe A, since a push on it is negative (toward the left).

5. Use the two carts to explore various situations that correspond to the predictions you made about interaction forces. Your goal is to find out under what circumstances one cart exerts more force on the other.

   Try collisions (a)–(c) listed below. If you have more time, try other collisions suggested in Extension 1-2.

   Be sure to **zero** the force probes before each collision. Also be sure that the forces during the collisions do not exceed 10 N.

   Sketch the graphs for each collision on the previous axes, or **print** them and affix them over the axes. Be sure to label your graphs.

   For each collision use the **integration routine** to find the values of the impulses exerted by each cart on the other (i.e., the areas under the force-time graphs). Record these values in the spaces below and carefully describe what you did and what you observed.

   (a) Two carts of the same mass moving toward each other at about the same speed.

   (b) Two carts of the same mass, one at rest and the other moving toward it.

   (c) One cart twice or three times as massive as the other, moving toward the other cart, which is at rest.

Question 1-1: Did your observations agree with your predictions? What can you conclude about forces of interaction during collisions? Under what circumstances does one object experience a different force than the other during a collision? How do forces compare on a moment by moment basis during each collision?

Question 1-2: You have probably studied *Newton's third law* in lecture or in your text. Do your conclusions have anything to do with *Newton's third law*? Explain.

Question 1-3: How does the impulse due to cart A acting on cart B compare to the impulse of cart B acting on cart A in each collision? Are they the same in magnitude or different? Do they have the same sign or different signs?

Extension 1-2: More Collision Interactive Forces

Make predictions for the interaction forces in the following situations, and then use the apparatus from Activity 1-1 to test your predictions. In each case describe your observations and how you made them. Include copies of any graphs you make. Compare your observations to your predictions.

Prediction E1-4: Suppose the mass of object A is greater than that of object B and the objects are moving toward each other at the same speed so

$$m_A > m_B \quad \text{and} \quad \vec{v}_B = -\vec{v}_A$$

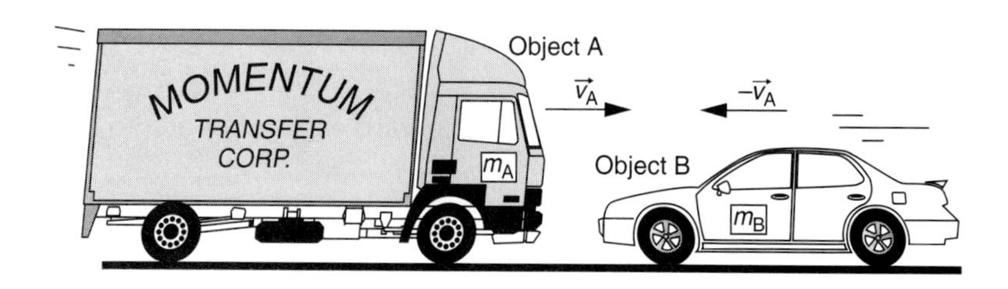

Predict the relative magnitudes of the forces between object A and object B.

______Object A exerts a larger force on object B.

______The objects exert the same size force on each other.

______Object B exerts a larger force on object A.

Prediction E1-5: Suppose the mass of object A is much greater than that of object B and that object B is moving in the same direction as object A but not as fast so

$$m_A \gg m_B \quad \text{and} \quad \vec{v}_B > \vec{v}_A$$

Predict the relative magnitudes of the forces between object A and object B.

______Object A exerts a larger force on object B.

______The objects exert the same size force on each other.

______Object B exerts a larger force on object A.

Prediction E1-6: Suppose the mass of object A is much greater than that of object B and that both objects are at rest until an explosion occurs so

$$m_A \gg m_B \quad \text{and} \quad \vec{v}_A = \vec{v}_B = 0$$

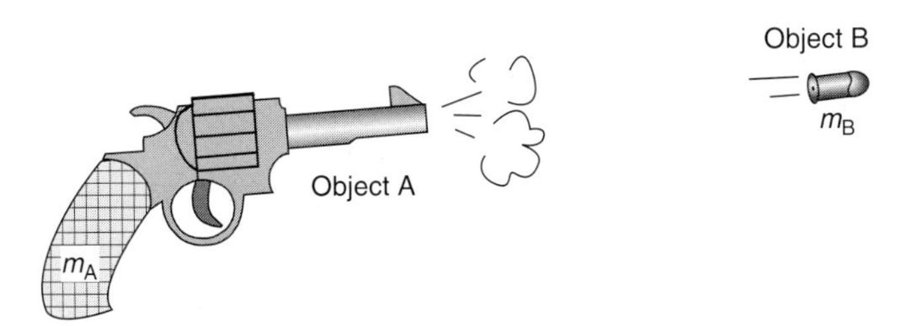

Predict the relative magnitudes of the forces between object A and object B. Place a check next to your prediction!

______Object A exerts a larger force on object B.

______The objects exert the same size force on each other.

______Object B exerts a larger force on object A.

Activity 1-3: Other Interaction Forces

Interaction forces between two objects occur in many other situations besides collisions. For example, suppose that a small car pushes a truck with a stalled engine, as shown in the picture. The mass of object A (the car) is much smaller than object B (the truck).

At first the car doesn't push hard enough to make the truck move. Then, as the driver pushes harder on the gas pedal, the truck begins to accelerate. Finally, the car and truck are moving along at the same constant speed.

Prediction 1-7: Place a check next to your predictions of the relative magnitudes of the forces between objects A and B.

Before the truck starts moving:

______the car exerts a larger force on the truck.

______the car and truck exert the same size force on each other.

______the truck exerts a larger force on the car.

While the truck is accelerating:

______the car exerts a larger force on the truck.

______the car and truck exert the same size force on each other.

______the truck exerts a larger force on the car.

After the car and truck are moving at a constant speed:

______the car exerts a larger force on the truck.

______the car and truck exert the same size force on each other.

______the truck exerts a larger force on the car.

Test your predictions.

1. Open the experiment file called **Other Interactions (L09A1-3)** to display the axes that follow. The software is now set up to display the two force probes at a slower data rate of 20 points per second.

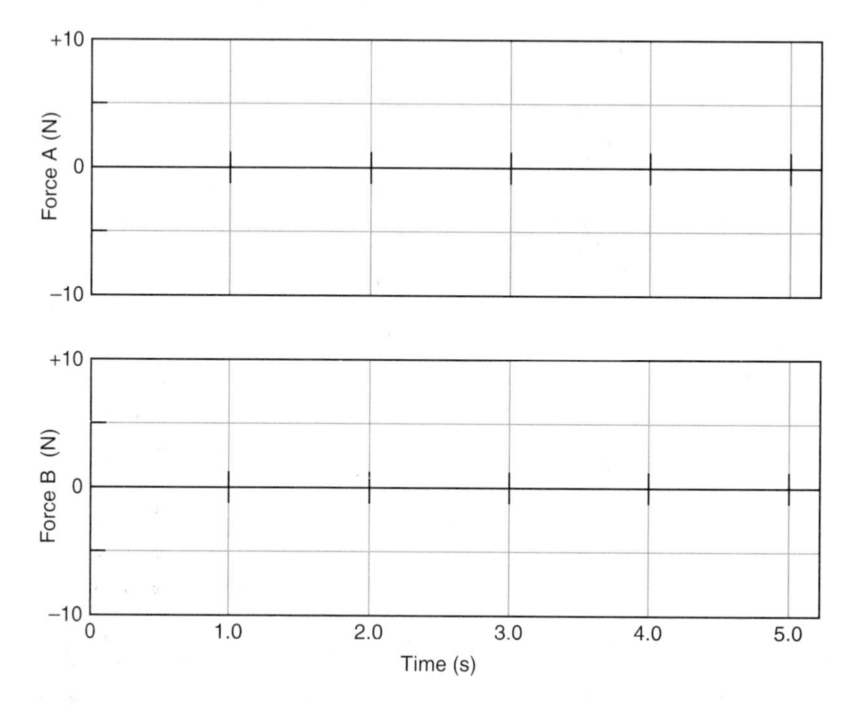

2. Use the same setup as in Activity 1-1 with the two force probes mounted on carts. Add masses to cart B (the truck) to make it much more massive than cart A (the car) (two or three times the mass).
3. **Zero** both force probes just before you are ready to take measurements.
4. Your hand will be the engine for cart A. Move the carts so that the stoppers are touching, and then **begin graphing.** When graphing begins, push cart A toward the right. At first hold cart B so it cannot move, but then allow the push of cart A to accelerate cart B, so that both carts move toward the right, finally at a constant velocity.
5. Sketch your graphs on the axes above, or **print** them and affix them over the axes.

Question 1-4: How do your results compare to your predictions? Is the force exerted by cart A on cart B (reading of force probe B) significantly different from the force exerted by cart B on cart A (reading of force probe A) during any part of the motion? Explain any differences you observe between your predictions and your observations.

Question 1-5: Explain how cart B is able to accelerate. Use *Newton's second law,* and analyze the combined (net) force exerted by all the forces acting on it. Is there a nonzero net force?

If you have additional time, do the following Extension.

Extension 1-4: More Interactions

Think of other physical situations with interaction forces between two objects that you can examine with the available two force probe apparatus. (For example, you might have one cart pushing the other as in Activity 1-3, but change the masses of the carts or change how they move. Or you might try an entirely new situation.)

Describe each physical situation completely. Draw a diagram. Predict the relative magnitudes of the two interaction forces. Then set up the apparatus and test your predictions. Describe your observations. Include copies of any graphs you make. Compare your observations to your predictions.

INVESTIGATION 2: NEWTON'S LAWS AND MOMENTUM CONSERVATION

Your previous work should have shown that interaction forces between two objects are equal in magnitude and opposite in sign (direction) on a moment by moment basis for all the interactions you might have studied. This is a testimonial to the seemingly universal applicability of *Newton's third law* to interactions between objects.

As a consequence of the forces being equal and opposite at each moment, you should have seen that the impulses of the two forces were always equal in magnitude and opposite in direction. This observation, along with the impulse–momentum theorem, is the basis for the derivation of the *conservation of momentum law,* which you may have seen in lecture or in your text. (The impulse–momentum theorem is really equivalent to *Newton's second law* since it can be derived mathematically from the *second law.*) The argument is that the impulse $\vec{J}_A$ acting on cart A during the collision equals the change in momentum of cart A, and the impulse $\vec{J}_B$ acting on cart B during the collision equals the change in momentum of cart B:

$$\vec{J}_A = \Delta\vec{p}_A \qquad \vec{J}_B = \Delta\vec{p}_B$$

But, as you have seen, if the only forces acting on the carts are the interaction forces between them, then $\vec{J}_A = -\vec{J}_B$. Thus, by simple algebra

$$\Delta\vec{p}_A = -\Delta\vec{p}_B \quad \text{or} \quad \Delta\vec{p}_A + \Delta\vec{p}_B = 0$$

i.e., there is no change in the total momentum $\vec{p}_A + \vec{p}_B$ of the system (the two carts).

If the momenta of the two carts before (initial—subscript i) and after (final—subscript f) the collision are represented in the diagrams below, then:

$$\vec{p}_f = \vec{p}_i$$

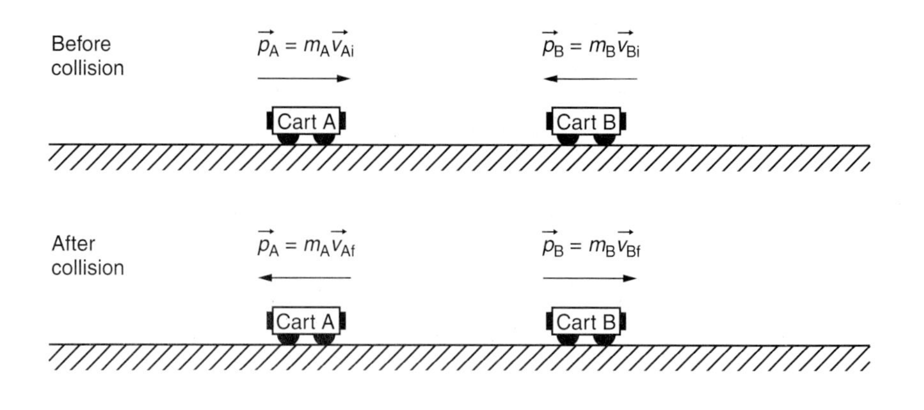

where

$$\vec{p}_i = m_A\vec{v}_{Ai} + m_B\vec{v}_{Bi} \qquad \vec{p}_f = m_A\vec{v}_{Af} + m_B\vec{v}_{Bf}$$

In the next activity you will examine whether momentum is conserved in a simple *inelastic* collision between two carts of unequal mass. You will need the following:

- computer-based laboratory system
- *RealTime Physics Mechanics* experiment configuration files
- two force probes with rubber stoppers replacing the hooks
- motion detector
- two low-friction carts
- masses to place on one of the carts to double its mass
- smooth level ramp or other level surface
- clay or Velcro

Activity 2-1: Conservation of Momentum in an Inelastic Collision

1. Set up the carts, ramp, and motion detector as shown below. Remove the force probes, and place blobs of clay or Velcro on the carts so that they will stick together after the collision. Add masses to cart A so that it is about twice as massive as cart B.

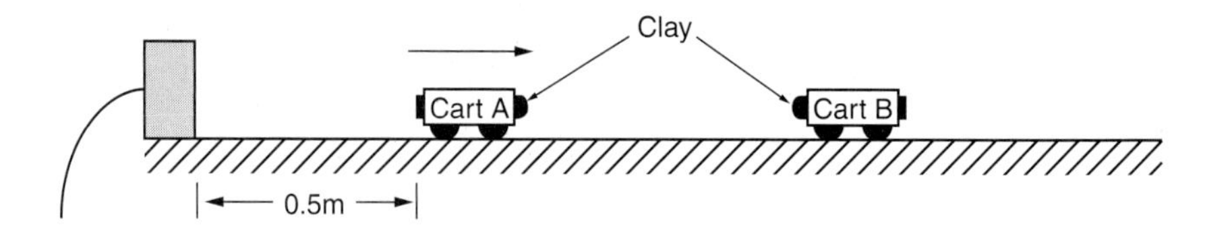

2. Measure the masses of the two carts.

$$m_A = \_\_\_\_\_\_\text{kg} \qquad m_B = \_\_\_\_\_\_\text{kg}$$

3. Open the experiment file called **Inelastic Collision (L09A2-1)** to display the axes that follow.

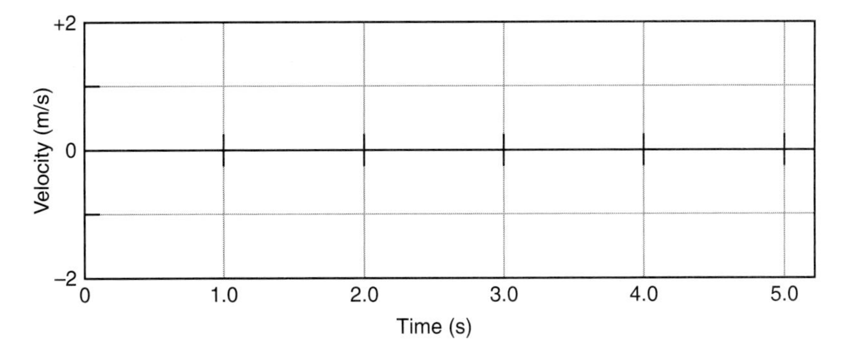

Prediction 2-1: You are going to give the more massive cart A a push and collide it with cart B, which is initially at rest. The carts will stick together after the collision. Suppose that you measure the *total* momentum of cart A and cart B before and after the collision. How do you think that the total momentum after the collision will compare to the total momentum before the collision? Explain the basis for your prediction.

4. Test your prediction. Begin with cart A at least 0.50 m from the motion detector. **Begin graphing,** and when you hear the clicks of the motion detector, give cart A a brisk push toward cart B and release it. *Be sure that the motion detector does not see your hand.*

   Repeat until you get a good run when the carts stick and move together after the collision. Then sketch the graph on the axes above, or **print** it and affix it over the axes.

5. Use the **analysis and statistics features** of the software to measure the velocity of cart A just before the collision and the velocity of the two carts together just after the collision. (You will want to find the average velocities over short time intervals just before and just after—but not during—the collision.)

   $v_{Ai} =$ ________m/s $\qquad v_{Af} = v_{Bf} =$ ________m/s

6. Calculate the total momentum of carts A and B before the collision and after the collision. Show your calculations below.

   $p_i =$ ________kg-m/s $\qquad p_f =$ ________kg-m/s

Question 2-1: Was momentum conserved during the collision? Did your results agree with your prediction? Explain.

Question 2-2: What are the difficulties in doing this experiment that might cause errors in the results?

Comment: Momentum is conserved whether or not the two carts stick to one another after the collision, but to test this you would need to measure the velocities of both carts after the collision. Momentum is also conserved if both carts are moving before the collision. If you have additional time, try to explore some of these other types of collisions in the following Extension.

Extension 2-2: Other Collisions

Consider other collisions that you can examine with the available apparatus. For example, you might change the masses of the two carts or have them both moving before the collision. (You might also try elastic collisions using the bumpers on the carts instead of clay, but for these it will be hard to measure more than one velocity with just one motion detector. It is possible to use two motion detectors, but this is somewhat more difficult to set up.)

Describe each physical situation completely. Draw a diagram. Predict how the total momentum after the collision will compare to that before the collision. Then set up the apparatus and test your predictions. Describe your observations. Include copies of any graphs you make. Compare your observations to your predictions.

Name________________________ Date____________ Partners________________________

HOMEWORK FOR LAB 9: NEWTON'S THIRD LAW AND CONSERVATION OF MOMENTUM

1. A weight lifter struggles but manages to keep a heavy barbell above his head. Occasionally he slips and the barbell starts to fall downward, but he always recovers. Compare the force exerted *by the weight lifter on the barbell* to that exerted *by the barbell on the weight lifter.* Is one force larger than the other or are they equal in magnitude to each other? Does this change during the times when the barbell slips downward? Explain.

2. A bowling ball rolls down the alley and hits a pin. Compare the force exerted *by the ball on the pin* to the force exerted *by the pin on the ball* during the collision between the ball and the pin. Is one force larger than the other or are they equal in magnitude to each other? Does this relationship of the magnitudes of the two forces change at any time during the collision? Explain.

3. A tennis player hits the ball with her racket. Compare the force exerted *by the racket on the ball* to that exerted *by the ball on the racket* during the collision between the racket and the ball. Is one force larger than the other or are they equal in magnitude to each other? Does this relationship of the magnitudes of the two forces change at any time during the collision? Explain.

4. Sketch on the axes on the right the force exerted *by a tennis racket on a ball* and the force exerted *by the ball on the racket* during a collision. Explain the shapes of your graphs.

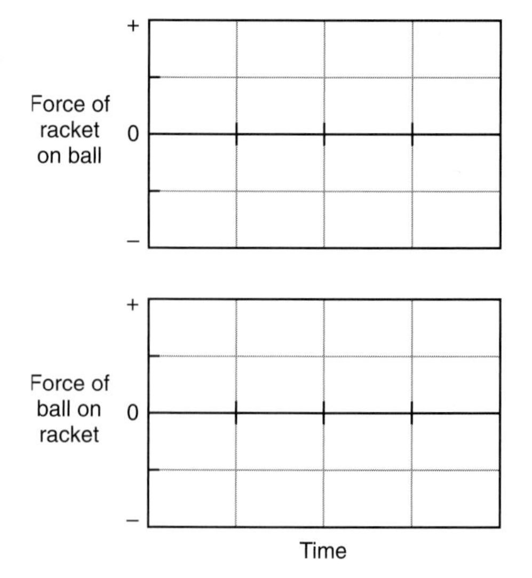

5. An automobile of mass 1500 kg moving at 25.0 m/s collides with a truck of mass 4500 kg at rest. The bumpers of the two vehicles lock together during the crash.

   a. Compare the force exerted *by the car on the truck* with that exerted *by the truck on the car* during the collision. Is one force larger than the other or are they equal in magnitude to each other?

   b. What is the final velocity of the car and truck just after the collision? Show your calculations.

6. The same collision as in Question 5 takes place, only this time the car and the truck bounce off each other *completely elastically.*

   a. Compare the force exerted *by the car on the truck* with that exerted *by the truck on the car* during the collision. Is one force larger than the other or are they equal in magnitude to each other?

   b. What are the final velocities of the car and truck just after the collision? Show your calculations.

Name______________________________ Date____________________

PRE-LAB PREPARATION SHEET FOR LAB 10: TWO-DIMENSIONAL MOTION (PROJECTILE MOTION)

(Due at the beginning of Lab 10)

Directions:
Read over Lab 10 and then answer the following questions about the procedures.

1. What are the equations for y and v_y in Question 1-1?

2. Sketch your answers to Question 1-4 on the axes below.

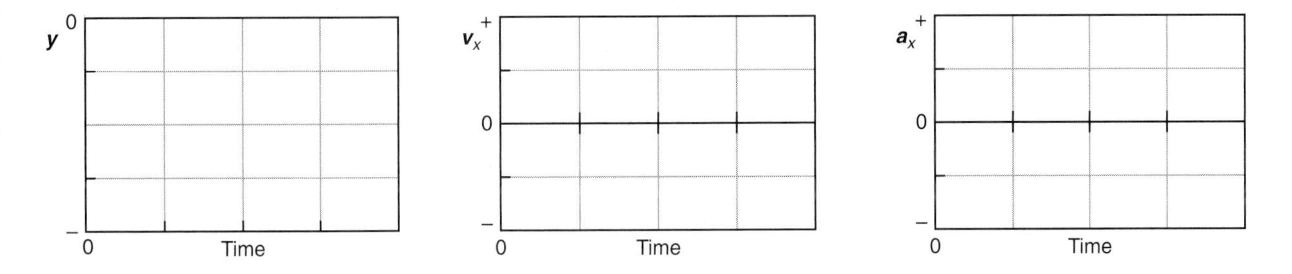

3. How will you calculate the average velocity, <v>, and average acceleration, <a>, in Table 1-1?

4. Sketch your Prediction 2-1 for the path followed by the ball rolling in the x direction and tapped in the y direction.

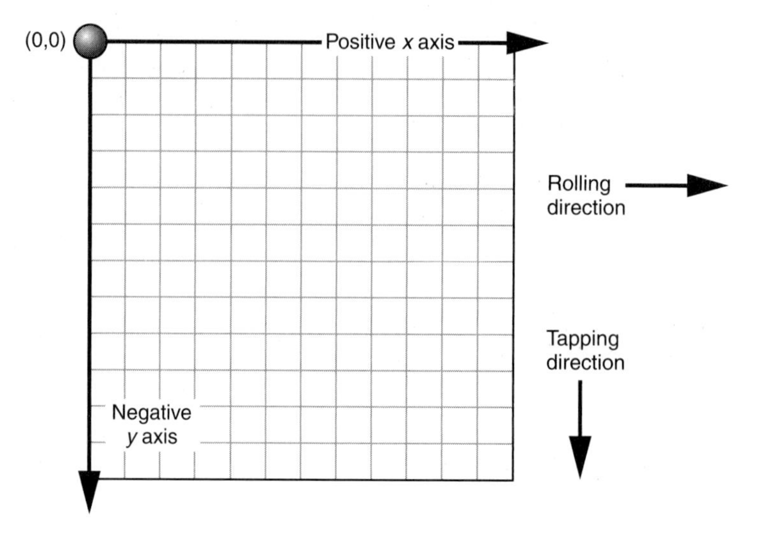

5. What graphs will you plot in Activity 2-1?

Name______________________ Date__________ Partners______________________

LAB 10: TWO-DIMENSIONAL MOTION (PROJECTILE MOTION)

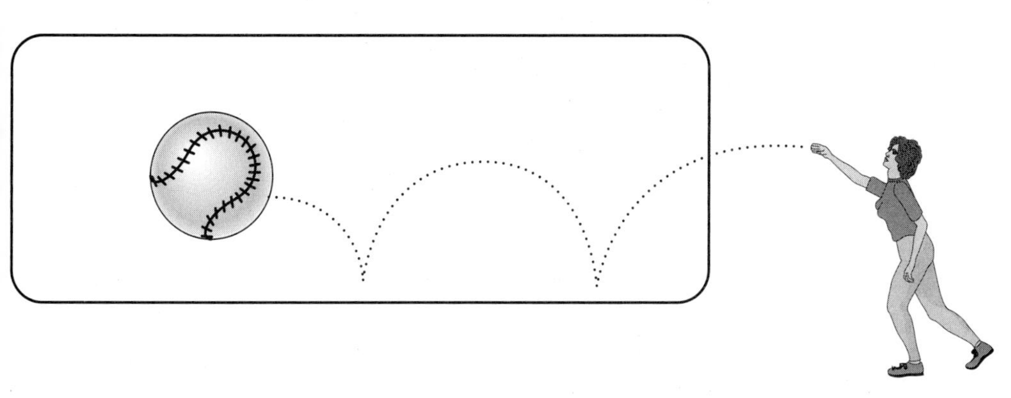

The essential fact is that all the pictures which science now draws of nature . . . are mathematical pictures.

—Sir James Jeans

OBJECTIVES

- To explore the similarity between the type of motion that results when an object falls close to the Earth's surface and that which results from accelerating a rolling ball by tapping it continuously.
- To review how vector quantities like velocities and accelerations in two dimensions can be represented as components that can be treated independently.
- To understand the experimental and theoretical basis for describing projectile motion as the superposition of two independent motions: (1) that of a body falling in the vertical direction under the influence of a constant force, and (2) that of a body moving in the horizontal direction with no applied forces.

OVERVIEW

So far we have been dealing separately with motion along a horizontal straight line and motion along a vertical straight line. The focus of this lab is the description of motion that occurs when an object is allowed to move in both the vertical and horizontal directions at the same time, close to the surface of the Earth. Examples are the motion of a baseball or tennis ball after it is hit, or the motion of a cannon ball. This type of motion is commonly called *projectile motion.* To understand this motion, it is helpful to study the vertical and the horizontal types of motion separately and then see how they might be combined.

This lab begins with a review of the classic *kinematic equations* that describe the relationships between instantaneous position, velocity, and acceleration for an object that moves along a straight line (called one-dimensional motion), which you have probably seen in lecture or in your textbook. You have already examined a number of examples of such motion. In some the object moved with a constant velocity (zero acceleration). In others—such as the motion of a cart pushed along by the constant force of a fan unit as in Lab 2 or the falling motion of a ball pulled by the constant gravitational force in Lab 6—the object moved with a constant acceleration (steadily increasing velocity).

In this lab you will consider the motion of a ball rolling on a level surface. You can apply a force by tapping it repeatedly with a rubber mallet.

This lab will be devoted to the creation and analysis of two-dimensional motion by tapping a ball with a rubber mallet. You will first see if tapping the ball repeatedly in its direction of motion produces a motion with a constant acceleration. Then you will roll the ball in one direction while tapping on it at right angles in the direction it was initially rolling in.

INVESTIGATION 1: MORE ABOUT ONE-DIMENSIONAL MOTION

So far you should have discovered that the vertical motion of a ball near the surface of the Earth has an acceleration that is essentially constant, caused by the nearly constant gravitational force. A standard set of equations can be derived—using the definitions of average and instantaneous velocity and acceleration—that describe motion of an object that undergoes constant acceleration. These equations are called the kinematic equations and they have probably been derived for you in lecture or in your textbook:

- $v_x = v_{0x} + a_x t$
- $x = x_0 + v_{0x} t + \frac{1}{2} a_x t^2$

Beware! These kinematic equations apply only when an object undergoes constant acceleration.

The symbols have the following meanings:

x = position along the x axis at time t.

v_x = instantaneous velocity along the x axis at time t.

a_x = constant acceleration along the x axis. (It could vary in time, but we have chosen to consider only those cases for which the acceleration is constant.)

x_0 = initial position at $t = 0$. (Tells you where the object was at the moment you started counting time.)

v_{0x} = initial velocity component along the x axis at $t = 0$. (Tells you how fast the object was moving at the moment you started counting time.)

Question 1-1: Suppose the object is moving in the y direction rather than the x direction, that the acceleration is a_y, that the initial y position coordinate is y_0, and that the initial velocity component, v_{0y}, is in the y direction also. How would you write the equations now? Show the new equations in the space below.

$v_y =$

$y =$

Question 1-2: Sketch on the axes below the position–time, velocity–time, and acceleration–time graphs for the motion described in Question 1-1.

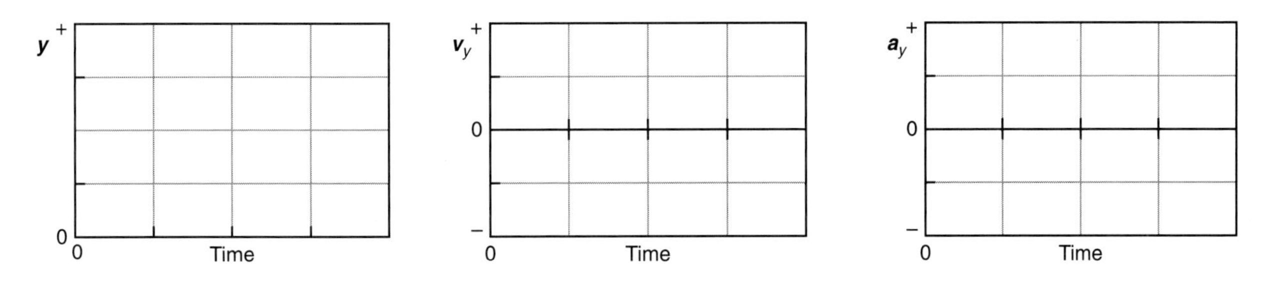

Question 1-3: Suppose that there is no acceleration in the x direction, that the initial x position coordinate is x_0, and that the initial x component of the velocity is v_{0x}. How would you write the equations for v_x and x as functions of time?

$v_x =$

$x =$

Question 1-4: Sketch on the axes below the position–time, velocity–time, and acceleration–time graphs for the motion of a bowling ball that is moving freely without being tapped. (Assume that friction can be neglected.)

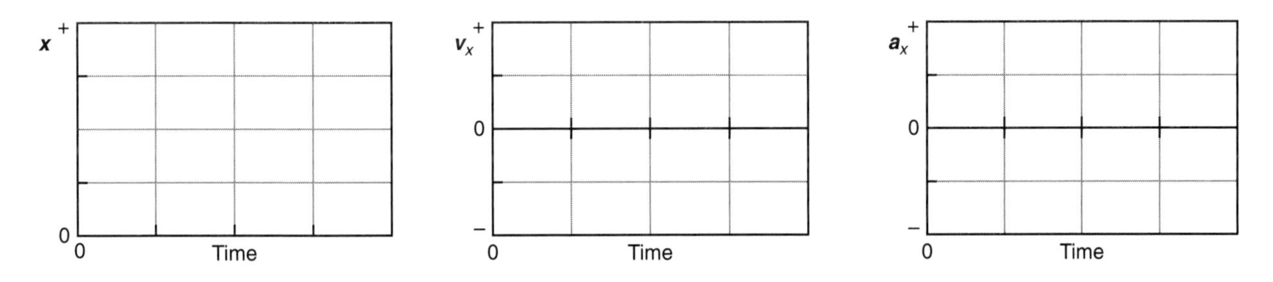

Can you learn how to tap a ball so that it will move with a constant acceleration? If so, you can use the similarity between the motion of a falling object and a tapped ball to study projectile motion in which an object falls vertically with a constant acceleration and moves horizontally with a constant velocity at the same time.

Let's start with one-dimensional measurements. For the measurements described below you will be using a twirling baton with a rubber tip to tap rapidly but gently on a duck pin ball (a small bowling ball) to produce an acceleration that is more or less constant. You will need

- duck pin ball (or regular bowling ball)
- twirling baton
- stopwatch
- tape measure or meter stick
- position markers (chalk, stickers, or sandbags)
- computer-based laboratory or data analysis software
- *RealTime Physics Mechanics* experiment configuration files

Activity 1-1: The Motion of a Tapped Ball

Find a stretch of fairly smooth level floor that is about 10 m long. A hallway is a good bet for this series of measurements. You are to record data for position vs. time for your ball starting at rest and receiving rapid, regular light taps. Before taking data, you and your partners should practice techniques for making these measurements.

(**Hints:** Concentrate on taking and recording data accurately. Practice coordinating tapping, timing, and marking positions several times before attempting to take data. Data that are not taken carefully will have too much variation to be interpreted. You should obtain at least 6 values of position.)

1. Set up a straight x axis on the floor, either using an existing line or a drawn one. Have one team member tap the ball, one be timer (who shouts out at the desired time instants), and one mark the position of the ball with chalk, stickers, or beanbags at the desired times. The ball should start from rest. The taps should be at a regular, rapid rate—*much more rapid than the rate at which you mark the ball's position.*

2. Choose an origin and measure positions (distances of the ball from the origin). Record these in Table 1-1. (Record data only in numbered rows. Add more rows if needed.)

3. Plot a graph of x position vs. time. You may open the experiment file called **Position vs. Time (L10A1-1),** and **enter the data** into the table provided. **Print** the graph and affix it below.

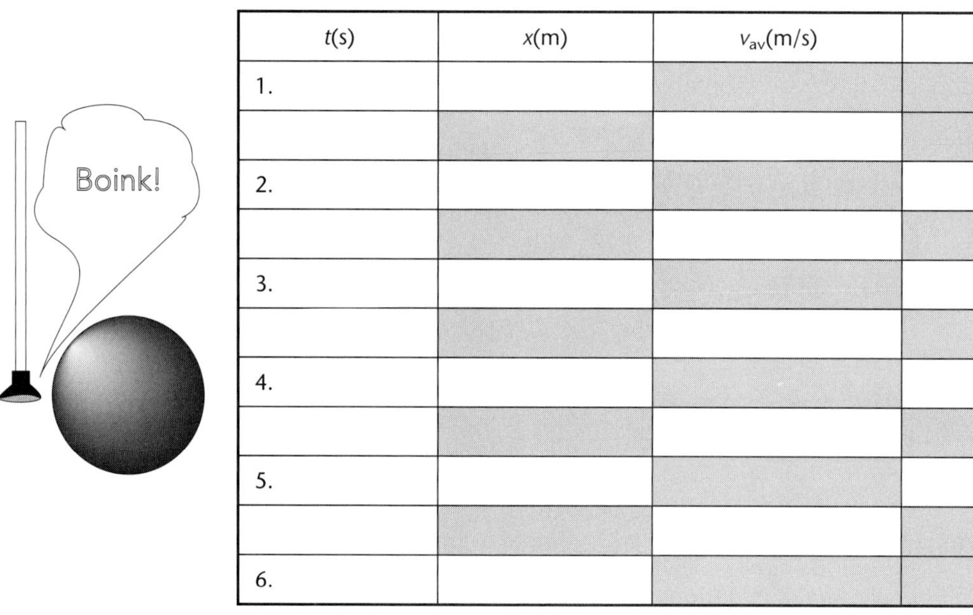

Table 1-1

	t(s)	x(m)	v_{av}(m/s)	a_{av}(m/s^2)
1.				
2.				
3.				
4.				
5.				
6.				

Question 1-5: Did your graph correspond to constant velocity motion (see Question 1-4) or constant acceleration motion (see Question 1-2)? Is this what you expected? Explain.

You can also use the data you just collected to plot a graph of velocity vs. time. Recall that the definitions of average velocity and average acceleration in one dimension are:

$$\langle v \rangle = \frac{\Delta x}{\Delta t}$$

$$\langle a \rangle = \frac{\Delta v}{\Delta t}$$

Activity 1-2: Velocity and Acceleration of a Tapped Ball

Once you have collected data in your table, analyze the data in the following way.

1. Calculate the average velocities in the time intervals between the measured positions in Table 1-1. For example, if the ball's position was marked every second, then you should calculate the average velocity in the time intervals 0.00 to 1.00 s, 1.00 to 2.00 s, and so on. Show a *sample* calculation clearly in the space below for the first time interval.

   In the interval from ______ to ______ s the average velocity is

   $\langle v \rangle =$

   This is equal to the velocity at the center of that time interval. (For example, if the first time interval were 0.00 to 1.00 s, then this would be the velocity at 0.50 s.)

2. Also calculate average accelerations from your calculated velocities. Show a *sample* calculation clearly in the space below for the first time interval.

   In the interval from ______ to ______ s the average acceleration is

   $\langle a \rangle =$

   (For example, if the two time intervals over which the velocities were calculated were 0.00 to 1.00 s and 1.00 to 2.00 s, then the two velocities would be at 0.50 and 1.50 s.)

3. Plot a graph of average velocity $\langle v \rangle$ vs. the time at the center of each time interval. (For the sample intervals above, the velocities would be plotted at 0.50 s, 1.50 s, etc.) You may open the experiment **Velocity vs. Time (L10A1-2),** and enter your data.

4. Use the **fit routine** in the software to fit a line to your velocity–time graph. **Print** your graph and affix it below.

Question 1-6: Is your velocity–time graph what you expected? Does it represent constant velocity or constant acceleration motion? Why does tapping the ball repeatedly result in this kind of motion?

Question 1-7: What was the magnitude of the acceleration of the ball from your graph? Explain how you found your answer.

Question 1-8: How does the acceleration compare to the average acceleration values calculated in your table? Is this what you expected?

INVESTIGATION 2: TWO-DIMENSIONAL VECTORS AND PROJECTILE MOTION

The world is full of phenomena that we can know of directly through our senses—objects moving, pushes and pulls, sights and sounds, winds and waterfalls. A vector is a mathematical concept—a mere figment of the mathematician's fancy. But vectors can be used to describe aspects of "real" phenomena such as positions, velocities, accelerations, and forces. Vectors are abstract entities invented by mathematicians that follow certain rules. For example, in the figure below, the velocity vector of an object is represented by the vector $\vec{\mathbf{v}}$, drawn relative to different coordinate axes.

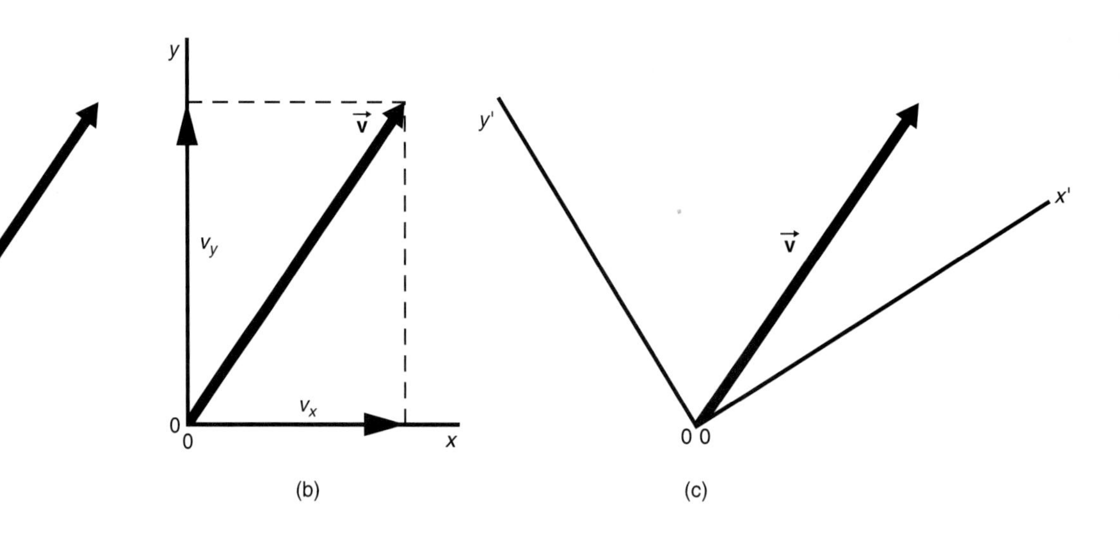

As discussed in Labs 1 and 2, a vector has two key attributes—magnitude and direction. The *magnitude* of a vector can be represented by the length of the arrow and its *direction* can be represented by the angle between the arrow and the coordinate axes chosen to help describe the vector.

In earlier labs you drew vectors in only one dimension. But vectors are especially useful in representing two-dimensional motion because they can be resolved into components. In the middle figure above, the velocity vector is resolved into x and y components. Added together, these components are equivalent to the original vector, but they can be analyzed *independently* of each other. This is one reason it is convenient to use vectors.

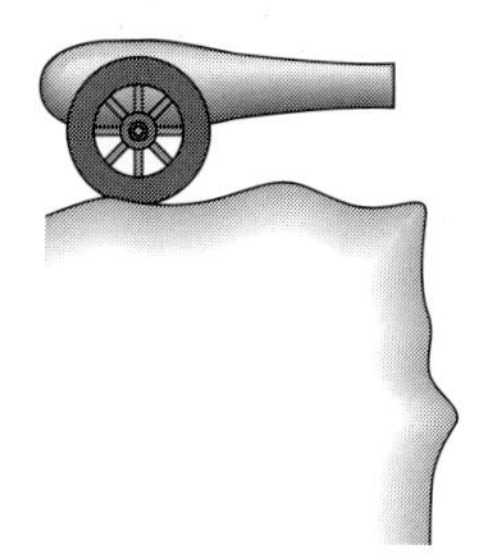

If a cannon ball is shot off a cliff with a certain initial velocity in the x direction, it will continue to move forward in that direction at the same velocity and at the same time fall in the y direction as a result of the gravitational attraction between the Earth and the ball. The two-dimensional motion that results is known as *projectile motion.*

In this lab you are going to simulate projectile motion by combining one-dimensional motion of the bowling ball at a constant velocity (x motion) with the constant acceleration motion that you have already observed (y motion). The constant velocity motion is that of the ball rolling along with no external forces on it. The constant acceleration motion is that of a ball receiving a series of rapid light taps. It is possible to combine these because the position, velocity, and acceleration of a projectile (and the ball) can be represented by vectors, and the x and y components of these vectors can be treated *independently.*

Once you have set up the motion, you can do a series of quantitative measurements to record the shape of the path described by the ball and measure its position, velocity, and acceleration in the x and y directions as the ball progresses.

Let's start with some predictions.

Prediction 2-1: Suppose you were to roll the ball briskly in one direction and then tap on it at right angles to its original direction. Can you guess what the resulting graph of its two-dimensional motion would look like? Sketch the predicted motion on the axes below. (**Hint:** Use your previous observations of constant acceleration motion in one dimension [e.g., ball falling or cart accelerated by the fan unit or falling mass] and of horizontal unaccelerated motion in one dimension to make an intelligent prediction of the path followed by the ball.)

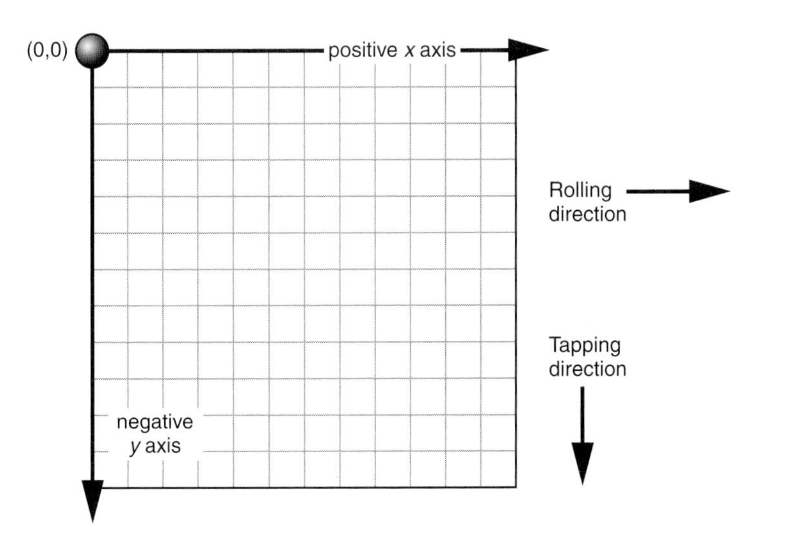

Explain the basis for your prediction.

Prediction 2-2: Sketch on the axes that follow your predictions for the position–time, velocity–time, and acceleration–time graphs in the x and y directions.

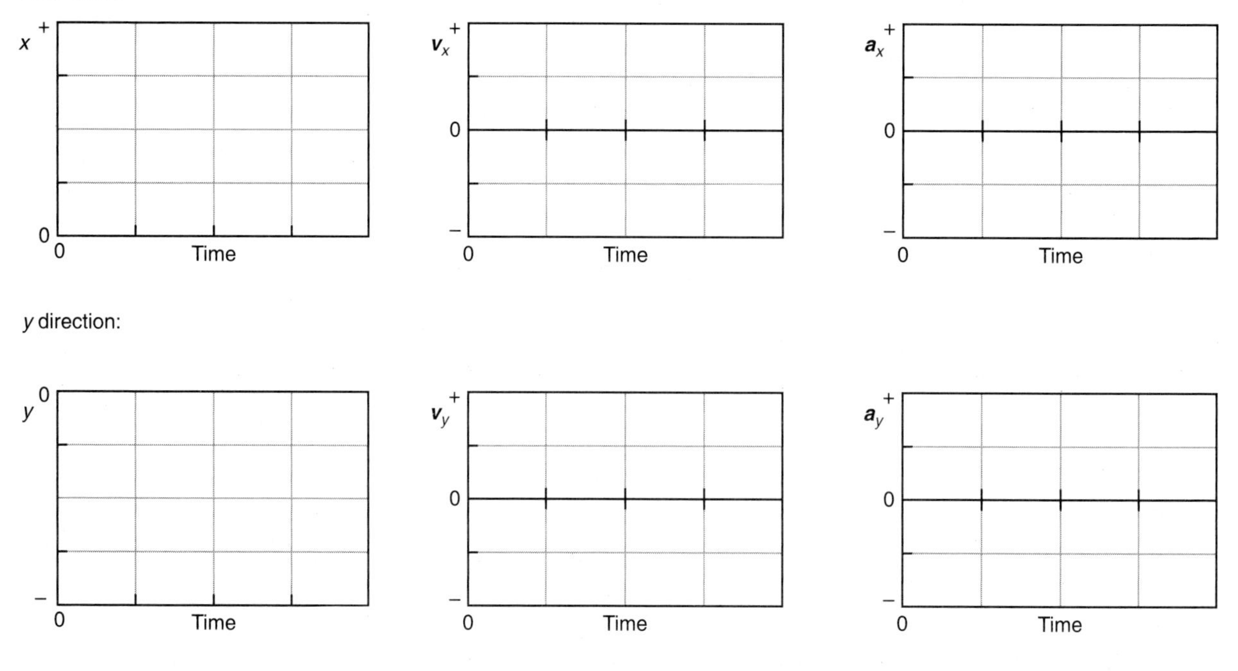

Test your predictions. To make the two-dimensional measurements described below you will be using the twirling baton and duck pin ball once again. You will need

- duck pin ball (or regular bowling ball) and twirling baton
- stopwatch
- tape measure or meter stick
- position markers (chalk, stickers, or small beanbags)
- computer-based laboratory or data analysis software
- *RealTime Physics Mechanics* experiment configuration files
- twirling baton

Activity 2-1: Determining the Path of a Ball in Two Dimensions

1. Find a stretch of fairly smooth, level floor that is about 10 m on a side. A gym floor is a good bet for this series of measurements. You are to record data for position vs. time for a ball that has a brisk initial velocity but is tapped lightly and regularly in the direction perpendicular (i.e., at right angles) to its *initial* velocity. Before taking data, you and your partners should practice techniques for making these measurements. Try to use the same tapping technique that you used in Activity 1-1. *No matter how the ball moves, be sure to tap only in the tapping direction (y direction).*

   (Hints: You will need to set up x and y axes and an origin, so that you can measure both the x and y coordinates of the ball. It will be helpful if you can use an already existing straight line, e.g., a wall, as one of your axes. Then, after you have rolled the ball and marked its position at fixed time intervals, its x and y coordinates can be measured as shown in the following figure.)

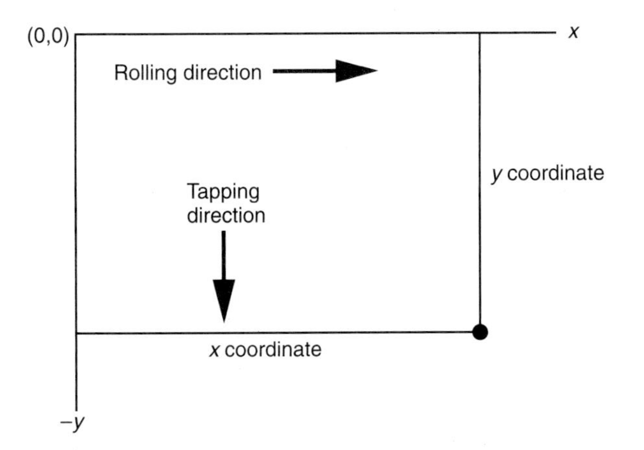

2. Measure the x and y positions of the ball relative to the axes as accurately as possible. Take your measurements along lines parallel to the x and y axes. Record your data as functions of time in Table 2-1 below. *Describe the procedures you used to take the data.*

3. Calculate $<v_x>$ and $<v_y>$ in each time interval, using the same methods as in Activity 1-1, and enter the values in the table.

Table 2-1

#	t (s)	x (m)	y (m)	$<v_x>$ (m/s)	$<v_y>$ (m/s)
1					
2					
3					
4					
5					
6					
7					
8					
9					
10					
11					
12					
13					
14					
15					
16					

4. Plot x vs. t and y vs. t using the experiment file called **2-D Position vs. Time (L10A2-1a). Print** the graphs and affix them in the space below. Be sure to label them.

Question 2-1: What kind of motion is the x motion—constant velocity (zero acceleration) or constant acceleration? Compare your answer to Prediction 2-2.

Question 2-2: What kind of motion is the y motion—constant velocity (zero acceleration) or constant acceleration? Compare your answer to Prediction 2-2.

5. **Enter the data** to plot $\langle v_x \rangle$ vs. t and $\langle v_y \rangle$ vs. t using the experiment file called **2-D Velocity vs. Time (L10A2-1b).**
6. Use the **fit routine** in the software to do a linear fit on each. **Print** your graphs and affix them below. Be sure to label them.

Question 2-3: Compare these graphs to Prediction 2-2. Do they agree with your predictions and your answers to Questions 2-1 and 2-2? Explain.

Question 2-4: Which motion—x or y—resembles free rolling motion and which resembles tapping starting from rest? Explain.

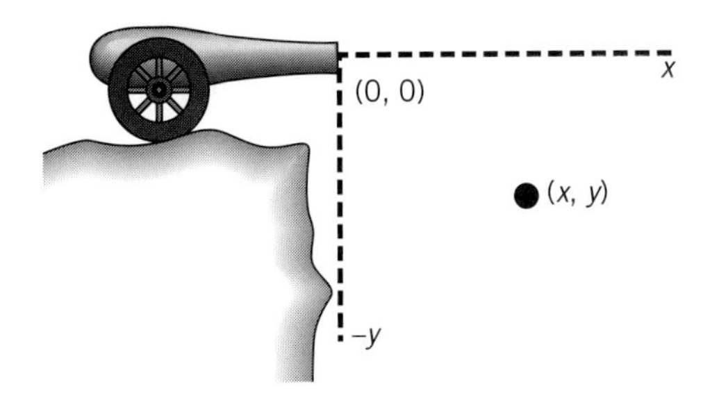

Question 2-5: A ball is shot from a cannon at the edge of a cliff, as shown on the left. Which direction of motion of the tapped rolling ball corresponds to the horizontal (x) motion of the cannon ball? Which direction of motion of the tapped rolling ball corresponds to the vertical (y) motion of the cannon ball? How do you know?

If you have time, do the following Extension to analyze your data more quantitatively.

Extension 2-2: More Analysis of the Ball's Motion

1. Find the numerical values of the x and y components of acceleration. (**Hint:** Use the **fit routine** in the software.) Record your values below along with the equations that fit the data.

   $a_x =$ equation:

   $a_y =$ equation:

Question E2-6: Are these values what you expect based on the types of motion in the x and y directions?

2. **Plot** a graph of y as a function of x. **Print** and affix the graph below.

Question E2-7: What is the shape of this graph? How does the graph compare to Prediction 2-1? How is the graph of y vs. x related to the two-dimensional path of the ball?

Question E2-8: Is tapping the ball a good simulation of projectile motion? Explain based on your results.

Name______________________________ Date______________ Partners______________________________

Homework for Lab 10: Two-Dimensional Motion (Projectile Motion)

1. A ball is released at the origin, and moves at a constant velocity on a *horizontal* floor along the dashed line shown below.

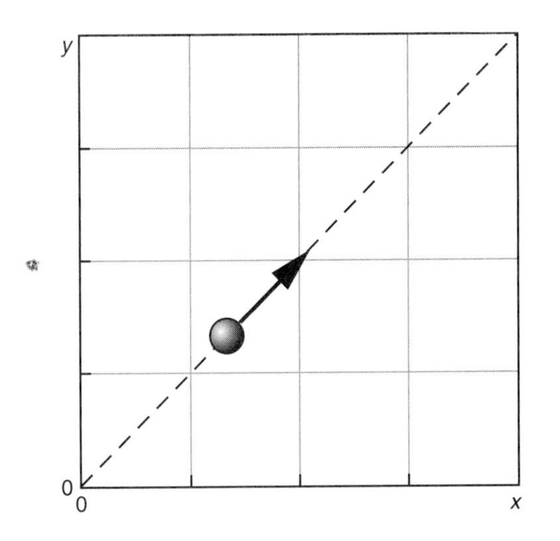

   a. Is this projectile motion? Explain your answer.

   b. Sketch on the axes that follow the x and y components of the position, velocity, and acceleration of the ball as functions of time.

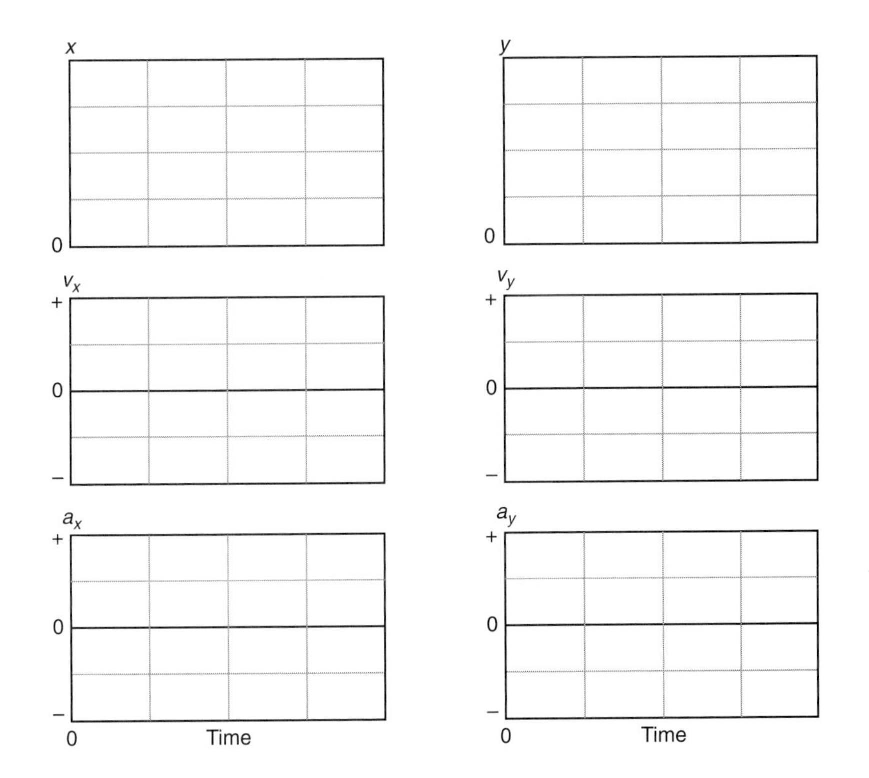

2. The ball in Question 1 begins *at rest* at the origin. It is tapped repeatedly by two tappers, with alternate taps in the x and y direction. It follows the same dashed path.

   a. Is this projectile motion? Explain your answer.

   b. Sketch on the axes that follow the x and y components of the position, velocity, and acceleration of the ball as functions of time.

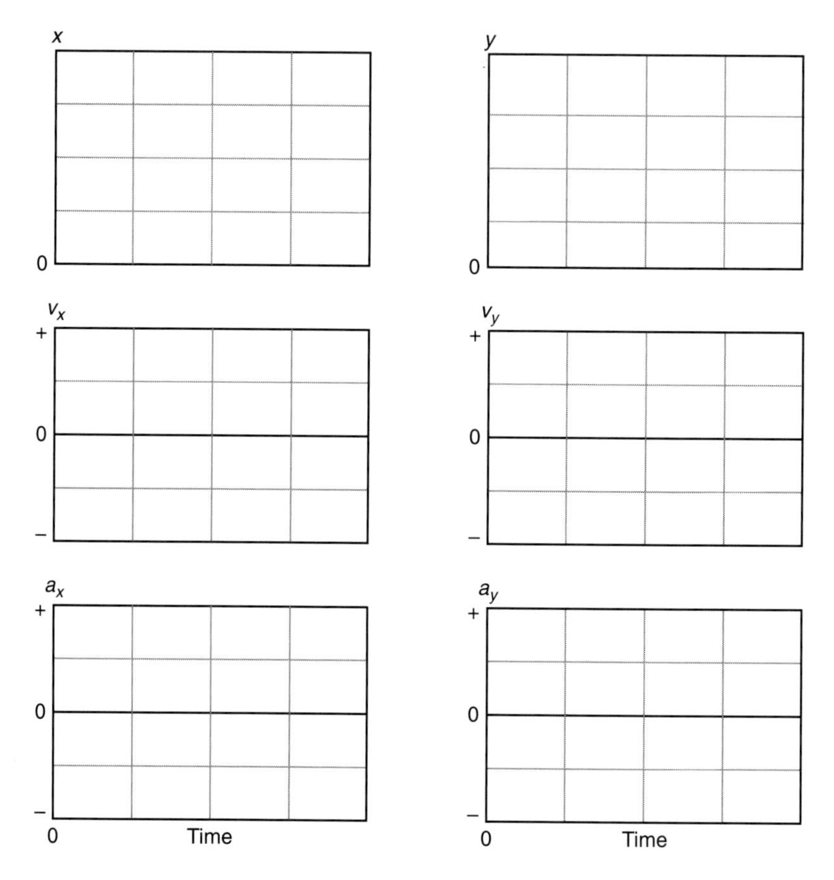

3. Explain why the motion of the ball that you observed in this lab when it rolled along while being tapped in the perpendicular direction is like projectile motion.

4. A pitcher throws a baseball. She releases it so that it is initially moving in the horizontal direction, as shown in the figure that follows. Is the motion of the ball after it leaves her hand projectile motion? Explain.

Assuming that air resistance is negligible, sketch the x and y components of the position, velocity, and acceleration of the ball as functions of time on the axes that follow. (Assume that she releases the ball at the origin of the coordinate system.)

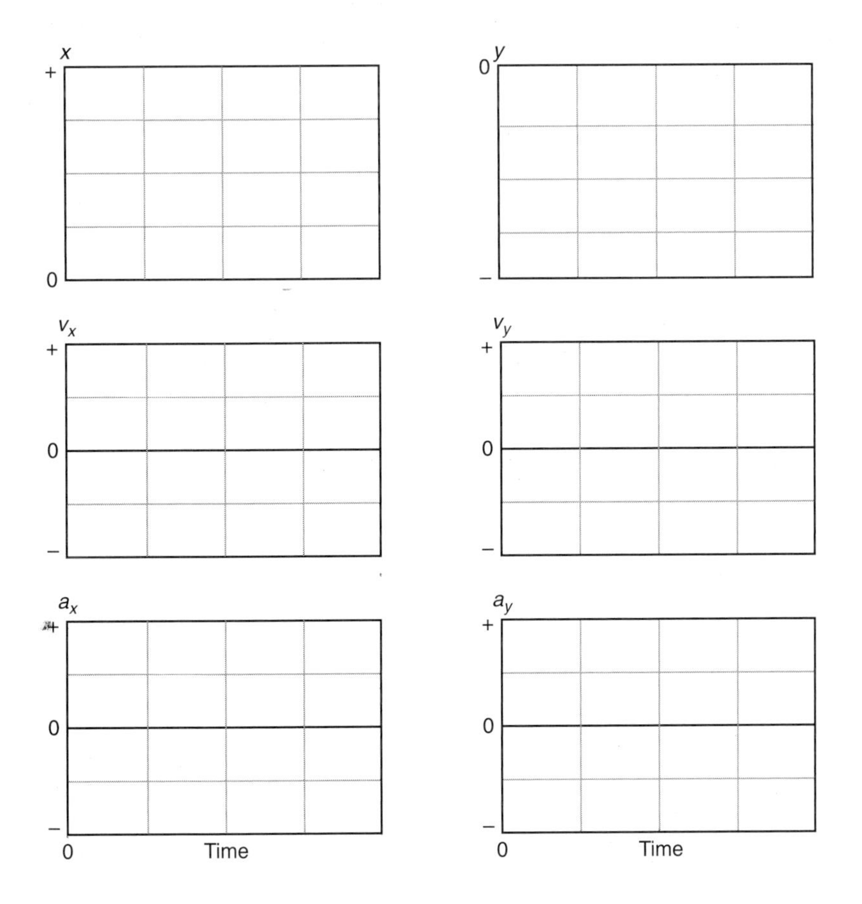

5. A ball is moving through the air. The data in the table that follows are either the x or y position coordinates of the ball as a function of time. (The positive y axis is upward.) Use these data to find velocities as a function of time, and

determine from the velocities whether these data are x or y coordinates. Be sure to explain thoroughly how you reached your conclusion. (You may wish to **plot a graph.**)

Time (s)	Position (m)
0.00	5.00
0.20	7.80
0.40	10.22
0.60	12.24
0.80	13.86
1.00	15.10
1.20	15.94
1.40	16.40
1.60	16.46
1.80	16.12
2.00	15.40

6. Suppose a rocket in outer space is thrust along the x direction with an acceleration of 15 m/s^2 while drifting freely (no applied force) in the y direction. Sketch below the path followed by the rocket.

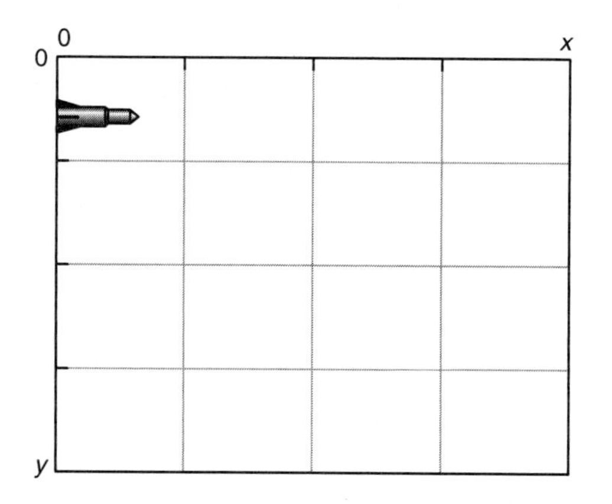

Is this motion similar to projectile motion? Explain.

Name______________________________ Date________________

PRE-LAB PREPARATION SHEET FOR LAB 11: WORK AND ENERGY

(Due at the beginning of Lab 11)

Directions:
Read over Lab 11 and then answer the following questions about the procedures.

1. What is your Prediction 1-1? Which job would you choose? Why?

2. In Activity 1-2, why are you asked to pull both parallel to the plane and at a 60° angle to the plane?

3. What is the definition of power?

4. How will you find the work done by the spring force in Activity 2-2?

5. How will work done and change in kinetic energy be compared in Activity 3-3?

Name______________________ Date____________ Partners______________________

LAB 11: WORK AND ENERGY

Energy is the only life and is from the Body; and Reason is the bound or outward circumference of energy. Energy is eternal delight.

—William Blake

OBJECTIVES

- To extend the intuitive notion of work as physical effort to a formal mathematical definition of work, W, as a function of both the force on an object and its displacement.
- To develop an understanding of how the work done on an object by a force can be measured.
- To understand the concept of power as the rate at which work is done.
- To understand the concept of kinetic energy and its relationship to the *net* work done on an object as embodied in the *work–energy principle.*

OVERVIEW

In your study of momentum in Labs 8 and 9, you saw that while momentum is always conserved in collisions, apparently different outcomes are possible. For example, if two identical carts moving at the same speed collide head-on and stick together, they both end up at rest immediately after the collision. If they bounce off each other instead, not only do both carts move apart at the same speed but in some cases they can move at the same speed they had coming into the collision. A third possibility is that the two carts can "explode" as a result of springs being released (or explosives!) and move faster after the interaction than before.

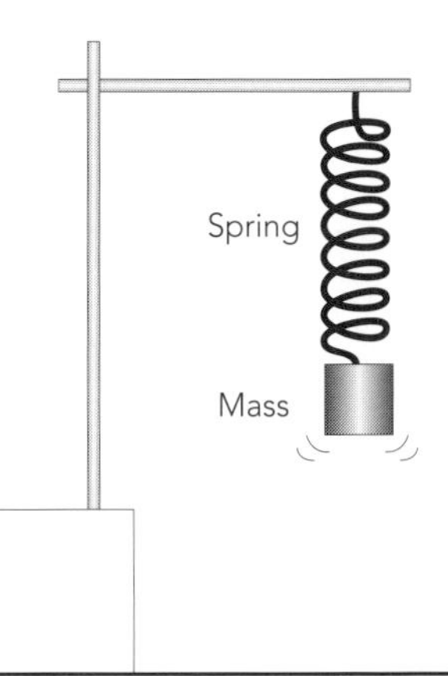

Two new concepts are useful in further studying various types of physical interactions—*work* and *energy.* In this lab, you will begin the process of understanding the scientific definitions of *work* and *energy,* which in some cases are different from the way these words are used in everyday language.

You will begin by comparing your intuitive, everyday understanding of work with its formal mathematical definition. You will first consider the work done on a small point-like object by a constant force. There are, however, many cases where the force is not constant. For example, the force exerted by a spring increases the more you stretch the spring. In this lab you will learn how to measure and calculate the work done by any force that acts on a moving object (even a force that changes with time).

Often it is useful to know both the total amount of work that is done, and also the rate at which it is done. The rate at which work is done is known as the *power.*

Energy (and the concept of conservation of energy, which we will explore in the next lab) is a powerful and useful concept in all the sciences. It is one of the more challenging concepts to understand. You will begin the study of energy in this lab by considering *kinetic energy*—a type of energy that depends on the velocity of an object and to its mass.

By comparing the change of an object's kinetic energy to the net work done on it, it is possible to understand the relationship between these two quantities in idealized situations. This relationship is known as the *work–energy principle.*

You will study a cart being pulled by the force applied by a spring. How much net work is done on the cart? What is the kinetic energy change of the cart? How is the change in kinetic energy related to the net work done on the cart by the spring?

INVESTIGATION 1: THE CONCEPTS OF PHYSICAL WORK AND POWER

While you all have an everyday understanding of the word "work" as being related to expending effort, the actual physical definition is very precise, and there are situations where this precise scientific definition does not agree with the everyday use of the word.

You will begin by looking at how to calculate the work done by constant forces, and then move on to consider forces that change with time.

Let's begin with a prediction that considers choosing among potential "real-life" jobs.

Prediction 1-1: Suppose you are president of the Load 'n' Go Company. A local college has three jobs it needs to have done and it will allow your company to choose one before offering the other two jobs to rival companies. All three jobs pay the same total amount of money.

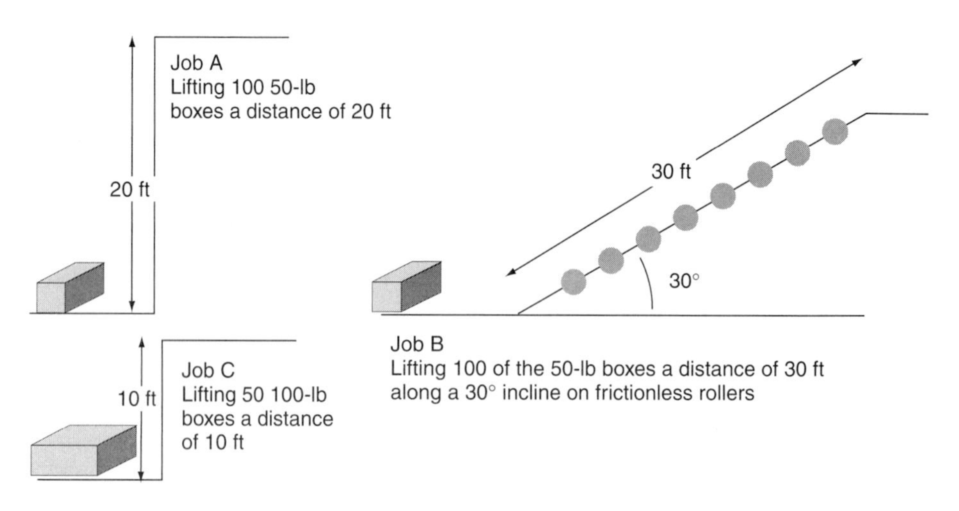

Which one would you choose for your crew? Explain why.

The following activities should help you to see whether your choice makes the most sense. You will need the following:

- computer-based laboratory system
- *RealTime Physics Mechanics* experiment configuration files
- force probe
- two bricks or two 1-kg masses
- low-friction cart
- smooth ramp or other level surface 2–3 m long which can be inclined
- meter stick
- protractor
- string
- 500-g mass

Activity 1-1: Effort and Work—Calculating Work

1. Lift a brick or a 1-kg mass at a slow, constant speed from the floor to a height of about 1 m. Repeat several times. Note the effort that is required.

   Repeat, this time lifting two bricks or two 1-kg masses 1 m.

2. Push a brick or a 1-kg mass 1 m along the floor at a constant speed. Repeat several times.

   Repeat, this time piling two bricks or two 1-kg masses on top of each other and pushing them 1 m.

Question 1-1: In each case, lifting or pushing, why must you exert a force to move the object?

Question 1-2: How much more effort does it take to lift or push two bricks or two 1-kg masses instead of one?

3. Lift a brick or a 1-kg mass with your hands at a slow, constant speed from the floor to a height of about 1 m. Repeat, this time lifting the brick or mass a distance of 2 m.

4. Push a brick or a 1-kg mass 1 m along the floor at a constant speed. Repeat, this time pushing the brick or mass a distance of 2 m.

Question 1-3: How much more effort does it take to lift or push an object twice the distance?

Question 1-4: If work were defined as "effort," how would you say work depends on the force applied and on the distance moved?

In physics, work is not simply effort. In fact, the physicist's definition of work is precise and mathematical. To have a full understanding of how work is defined in physics, we need to consider its definition for a very simple situation and then enrich it later to include more realistic situations.

> **Note:** All of the definitions of work in this unit apply only to very simple objects that can be idealized as point masses or are essentially rigid objects that don't deform appreciably when acted on by a force. The reason for limiting the definition to such objects is to avoid considering forces that cause the shape of an object to change or cause it to spin instead of changing the velocity or position of its center of mass.

If a rigid object or point mass experiences a constant force along the same line as its motion, the *work* done by that force is defined as the product of the force and the displacement of the center of mass of the object. Thus, in this simple situation where the force and displacement lie along the same line

$$W = F_x \, \Delta x$$

where W represents the work done by the force, F_x is the force, and Δx is the displacement of the center of mass of the object along the x axis. Note that if the force and displacement (direction of motion) are in the same direction (i.e., both positive or both negative), the *work done by the force is positive.* On the other hand, a force acting in a direction opposite to displacement does *negative work.* For example, an opposing force that is acting to slow down a moving object is doing *negative work.*

Question 1-5: Does this definition of work agree with the amount of effort you had to expend when you moved bricks or 1-kg masses under different conditions? Explain.

Question 1-6: Does effort necessarily result in physical work? Suppose two people are in an evenly matched tug of war. They are obviously expending *effort* to pull on the rope, but according to the definition are they doing any *physical work* as defined above? Explain.

Activity 1-2: Calculating Work When the Force and Displacement Lie Along the Same Line and When They Don't

In this activity, you will measure the force needed to pull a cart up an inclined ramp using a force probe. You will examine two situations. First you will exert a force parallel to the surface of the ramp, and then you will exert a force at an angle to the ramp. You will then be able to see how to calculate the work when the force and displacement are not in the same direction in such a way that the result makes physical sense.

1. Open the experiment file called **Force and Work (L11A1-2).** This will enable you to display only force data on axes like those shown below for a time interval of 10 s.

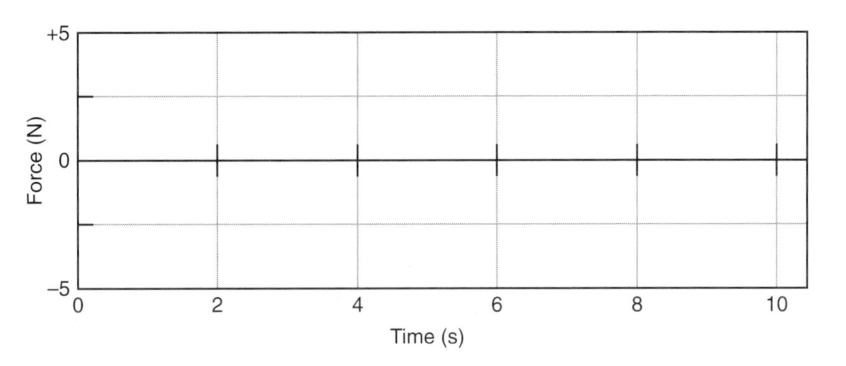

2. **Calibrate** the force probe with a 500-g mass (4.9-N weight) or **load the calibration.** (If you are using a Hall effect force probe, you may need to adjust the spacing and **check the sensitivity.**)

3. Set up the force probe, cart, and ramp as shown in the diagram below. Attach a short string (about 15 cm) to the front of the cart and make a loop in its end for the force probe. Support one end of the ramp so that it is inclined to an angle of about 15–20°.

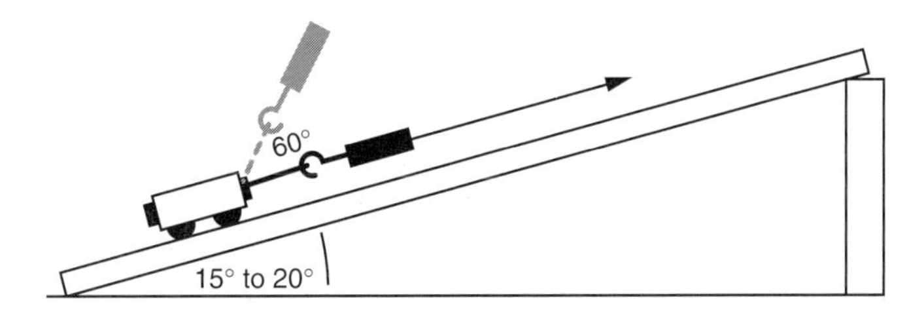

Reminder: Since the zero point of the force probe can change somewhat, it is important to always **zero** the probe (especially a Hall effect probe) with nothing pushing or pulling on the hook just before making any measurements.

4. Find the force needed to pull the cart up the ramp at a constant velocity. Hook the force probe through the loop in the string. **Zero** the force probe without pulling on the string. **Begin graphing** force vs. time as you pull the cart up the ramp slowly *at a constant velocity.* Pull the cart so that the string is always *parallel to the ramp.* Pull the cart a measured distance along the ramp, say 1.5 m.

5. Move your data so that the graph remains **persistently displayed on the screen** for comparison to the next graph. Then sketch your graph on the previous axes, or **print** and affix over the axes.

Prediction 1-2: Suppose that the force is not exerted along the line of motion but is in some other direction. If you try to pull the cart up along the same ramp in the same way as before (again with a constant velocity), only this time with a force that is not parallel to the surface of the ramp, will the force probe measure the same force, a larger force, or a smaller force?

Now test your prediction by measuring the force needed to pull the cart up along the ramp at a constant velocity, pulling at an angle of 60° to the surface of the ramp.

6. **Zero** the force probe. Attach it to the loop in the string as before. Measure the 60° angle with a protractor. **Begin graphing** force as you pull the cart up *at a slow constant speed* as shown in the diagram above. *Be sure the cart does not lift off the surface of the ramp.*
7. Sketch your graph on the previous axes, or **print** and affix over the axes.
8. Use the **analysis and statistics features** in the software to find the average (mean) force applied to the cart in both cases during the time intervals when the cart was moving with a constant velocity. Do not include the force to get the cart moving.

   Average force pulling parallel to surface:______N

   Average force pulling at 60° to the surface:______N

Question 1-7: Was the average force measured by the force probe different when the cart was pulled at 60° to the surface than when the cart was pulled parallel to the surface? Did the result agree with your prediction?

Question 1-8: Did it seem to take more effort to move the cart when the force was inclined at an angle to the ramp's surface? Do you think that more physical work was done to move the cart over the same distance at the same slow constant speed?

It is the force component parallel to the displacement that is included in the calculation of work. Thus, when the force and displacement are not parallel, the work is calculated by

$$W = F_x \, \Delta x = (F \cos \theta)\Delta x$$

Question 1-9: Do your observations support this equation as a reasonable way to calculate the work? Explain.

Question 1-10: Based on all of your observations in this investigation, was your choice in Prediction 1-1 the best one? In other words, did you pick the job requiring the least physical work? Explain.

Sometimes more than just the total physical work done is of interest. Often what is more important is the rate at which physical work is done. Average power, $<P>$, is defined as the ratio of the amount of work done, ΔW, to the time interval, Δt, in which it is done, so that

$$<P> = \frac{\Delta W}{\Delta t}$$

If work is measured in joules and time in seconds then the fundamental unit of power is the joule/second, and one joule/second is defined as one watt.

A more traditional unit of power is the horsepower, which originally represented the rate at which a typical work horse could do physical work. It turns out that

1 horsepower (or hp) = 746 watts

Those of you who are car buffs know that horsepower is used to rate high-performance cars. The engine in a high-performance car can produce hundreds of horsepower.

How does your lifting ability stack up? If you have time, do the following Extension.

Extension 1-3: Your Personal Power Rating

Let's see how long it takes you or one of your classmates to lift a heavy object like a bowling ball a distance of 1 m. For this observation you'll need the following:

- meter stick
- bowling ball
- digital stopwatch

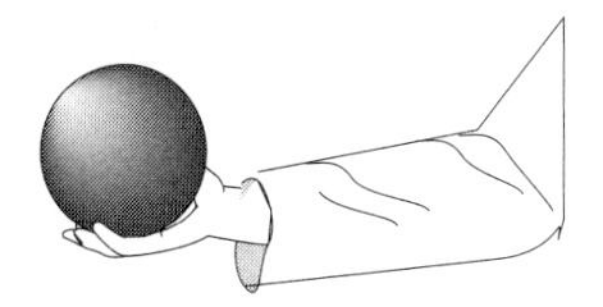

1. Lift a bowling ball through a known height as fast as possible. Measure the mass of the ball, the height of the lift, and the time.

   Mass______kg   Height______m   Time______s

2. Compute the work done against the force of gravity.

3. Compute the average power $<P>$ you expended in watts and in horsepower.

Question E1-11: How does your horsepower compare to that of your favorite automobile? If you're not interested in cars, how do *you* stack up against a horse?

INVESTIGATION 2: WORK DONE BY CONSTANT AND NONCONSTANT FORCES

Many forces in nature are not constant. A good example is the force exerted by a spring as you stretch it. In this investigation you will see how to calculate work and power when a nonconstant force acts on an object.

You will start by looking at a somewhat different way of calculating the work done by a constant force by using the area under a graph of force vs. position. It turns out that, unlike the equations we have written down so far, which are only valid for constant forces, the method of finding the area under the graph will work for both constant and changing forces.

You will need the following:

- computer-based laboratory system
- *RealTime Physics Mechanics* experiment configuration files
- force probe
- motion detector
- rod support for force probe
- 200-g and 500-g masses
- low-friction cart and flag
- smooth ramp or other level surface 2–3 m long
- meter stick
- spring

Activity 2-1: Work Done by a Constant Lifting Force

In this activity you will measure the work done when you lift an object from the floor through a measured distance. You will use the force probe to measure the force, and the motion detector to measure distance.

1. The motion detector should be on the floor, pointing upward.

2. **Calibrate** the force probe with a hanging mass of 500 g (force of 4.9 N) or **load the calibration** (if it hasn't already been calibrated in Investigation 1.) (If you are using a Hall effect force probe, the spacing between the magnet and the sensor should be the same as in Investigation 1.)

3. Open the experiment file called **Work in Lifting (L11A2-1).** This will allow you to display velocity and force for 10 s on the axes that follow.

4. **Zero** the force probe with the hook pointing vertically downward. Then hang a 200-g mass from its end, and **begin graphing** while lifting the mass at a slow, constant speed through a distance of about 1.0 m starting at least 0.5 m above the motion detector.

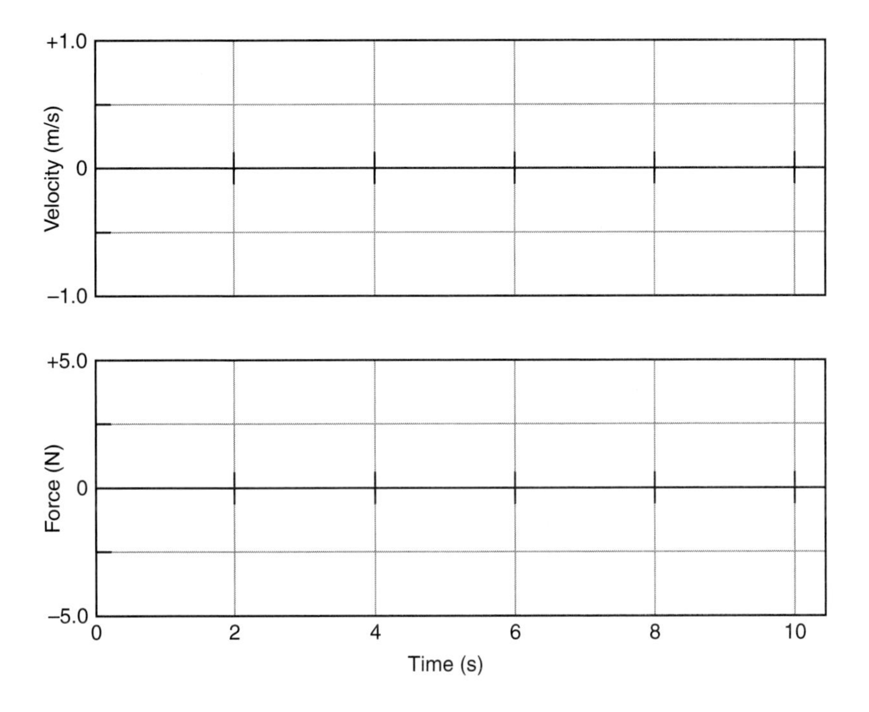

5. When you have a set of graphs in which the mass was moving at a reasonably constant speed, sketch your graphs on the axes, or **print** and affix over the axes.

Question 2-1: Did the force needed to move the mass depend on how high it was off the floor, or was it reasonably constant?

6. **Change to a force vs. position graph.** Sketch or **print** the graph below. (If sketched, fill in the scale on the position axis.)

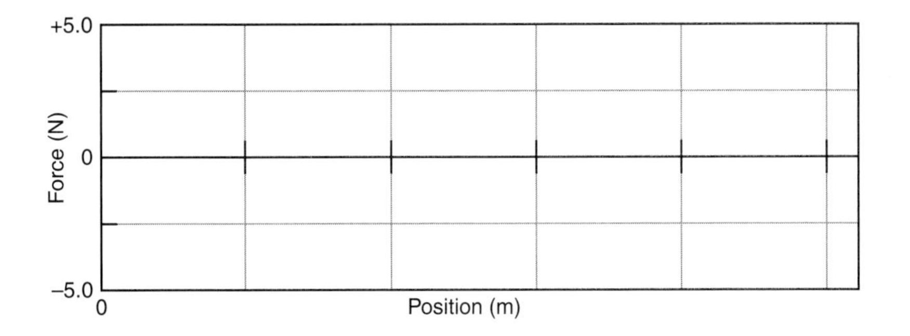

7. Use the **analysis and statistics features** of the software to find the average force over the distance the mass was lifted. Record this force and distance below.

   Average force:______N   Distance lifted:______m

8. Calculate the work done in lifting the mass. Show your calculation.

   Work done:______ J

9. Notice that force times distance is also the area of the rectangle under the force vs. position graph. Find the area under the curve directly by using the **integration routine** in the software.

   Area under force vs. position graph:______ J

Question 2-2: Do the two calculations of the work seem to agree with each other? Explain.

> **Comment:** This activity has dealt with the constant force required to lift an object against the gravitational force at a constant speed. The area under the force vs. position curve always gives the correct value for work, even when the force is not constant. (If you have studied calculus you may have noticed that the method of calculating the work by finding the area under the force vs. position graph is the same as integrating the force with respect to position.)

If you have time, find the work done by the nonconstant spring force using this method in Extension 2-2.

Extension 2-2: Work Done by a Nonconstant Spring Force

In this extension you will measure the work done when you stretch a spring through a measured distance. First you will collect data for force applied by a stretched spring vs. distance the spring is stretched, and you will plot a graph of force vs. distance. Then, as in Activity 2-1, you will be able to calculate the work done by finding the area under this graph.

1. Set up the ramp, cart, flag, motion detector, force probe, and spring as shown in the diagram.

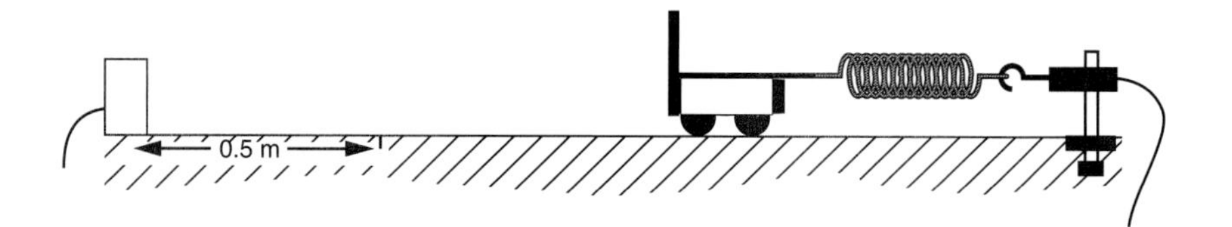

2. **Calibrate** the force probe with a force of 4.9 N applied to it or **load the calibration** (if it hasn't already been calibrated in Investigation 1.) (If you are using a Hall effect force probe, the spacing between the magnet and the sensor should be the same as in Investigation 1.)

Comment: We assume that the force measured by the force probe is the same as the force applied by the cart to the end of the spring. This is a consequence of *Newton's third law.*

3. *Be sure that the motion detector sees the cart over the whole distance of interest—from the position where the spring is just unstretched to the position where it is stretched about 1.0 m.*

4. Open the experiment file called **Stretching Spring (L11E2-2)** to display the force vs. position axes that follow.

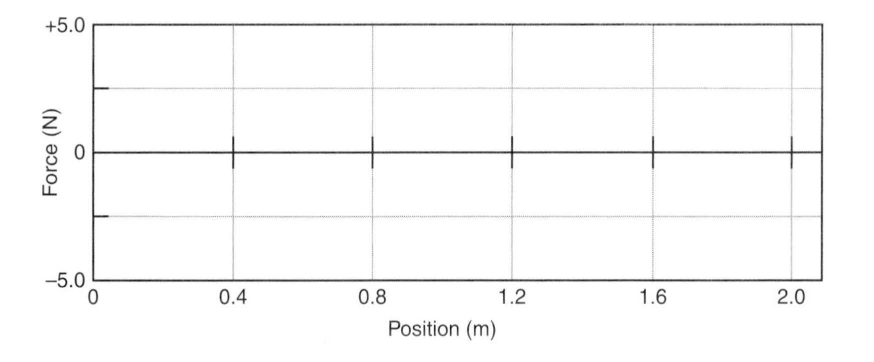

5. **Zero** the force probe with the spring hanging loosely. Then **begin graphing** force vs. position as the cart is moved slowly towards the motion detector until the spring is stretched about 1.0 m. (Keep your hand out of the way of the motion detector.)

6. Sketch your graph or **print** and affix over the axes.

Question E2-3: Compare this force vs. position graph to the one you got lifting the mass in Activity 2-1. Is the spring force a constant force? Describe any changes in the force as the spring is stretched.

Question E2-4: Can you use the equation $W = F_x \Delta x$ for calculating the work done by a nonconstant force like that produced by a spring? Explain.

7. Use the **integration routine** in the software to find the work done in stretching the spring.

Area under force vs. position graph:______ J

Investigation 3 will begin with an exploration of the definition of *kinetic energy*. Later, we will return to this method of measuring the area under the force vs. position graph to find the work, and we will compare the work done to changes in the kinetic energy.

INVESTIGATION 3: KINETIC ENERGY AND THE WORK–ENERGY PRINCIPLE

What happens when you apply an external force to an object that is free to move and has no frictional forces on it? According to *Newton's second law,* it should experience an acceleration and end up moving with a different velocity. Can we relate the change in velocity of the object to the amount of work that is done on it?

Consider a fairly simple situation. Suppose an object is lifted through a distance and then allowed to fall near the surface of the Earth. During the time it is falling it will experience a constant force as a result of the attraction between the object and the Earth—glibly called gravity or the force of gravity. You discovered how to find the work done by this force in Investigations 1 and 2. It is useful to define a new quantity called *kinetic energy.* You will see that as the object falls, its kinetic energy increases as a result of the work done by the gravitational force, and that, in fact, it increases by an amount exactly equal to the work done.

First you need to find a reasonable definition for the *kinetic energy.* You will need the following:

- rubber ball
- rubber ball with about twice the mass

Activity 3-1: Kinetic Energy

In this activity you will explore the meaning of kinetic energy, and see how it is calculated.

1. Go outside or out in the hall and toss the less massive rubber ball back and forth, slowly at first and then faster. Then alternate between throwing slowly and faster. (*Don't throw it so quickly that your partner is uncomfortable—this is not a contest!*) Notice how much effort it takes to throw it and to catch (stop) it when it is moving quickly or slowly.

Question 3-1: Does the effort needed to stop the ball seem to change as its speed increases? How does it change? Explain.

Question 3-2: Does the effort needed to throw the ball seem to change as its speed increases? How does it change? Explain.

2. Now throw the more massive ball. Toss the ball back and forth at the same speed. Then alternate between the heavier and lighter ball. Again notice how much effort it takes to throw and stop the ball.

Question 3-3: Does the effort needed to stop the ball seem to change as its mass increases? How does it change? Explain.

Question 3-4: Does the effort needed to throw the ball seem to change as its mass increases? How does it change? Explain.

Comment: When an object moves, it possesses a form of energy because of the work that was done to start it moving. This energy is called *kinetic energy.* You should have discovered that the amount of kinetic energy increases with both mass and speed. In fact, the kinetic energy is defined as being proportional to the mass and the square of the speed. The mathematical formula is

$$K = \frac{1}{2} mv^2$$

The unit of kinetic energy is the joule (J), the same as the unit of work.

Activity 3-2: Your Kinetic Energy

In this activity you will examine how you can graph the kinetic energy of an object such as your body in real time. You will need the following:

- computer-based laboratory system
- *RealTime Physics Mechanics* experiment configuration files
- motion detector

1. Open the experiment file **Kinetic Energy (L11A3-2).** This will display velocity vs. time axes like the ones that follow.
2. To display kinetic energy you will need to know your mass in kilograms. Use the fact that 1.0 kg weighs 2.2 lb on Earth to find your mass in kilograms.

   Mass:______kg

3. **Configure the software with a new column** calculated from one-half of your mass times the square of the velocity measured by the motion detector. Then both velocity and kinetic energy will be graphed in real time.
4. You are ready to record your velocity and kinetic energy as you walk. **Begin graphing** while walking away from the motion detector slowly, then more quickly, and then back toward the motion detector slowly and then more quickly. Sketch or **print** your graphs.

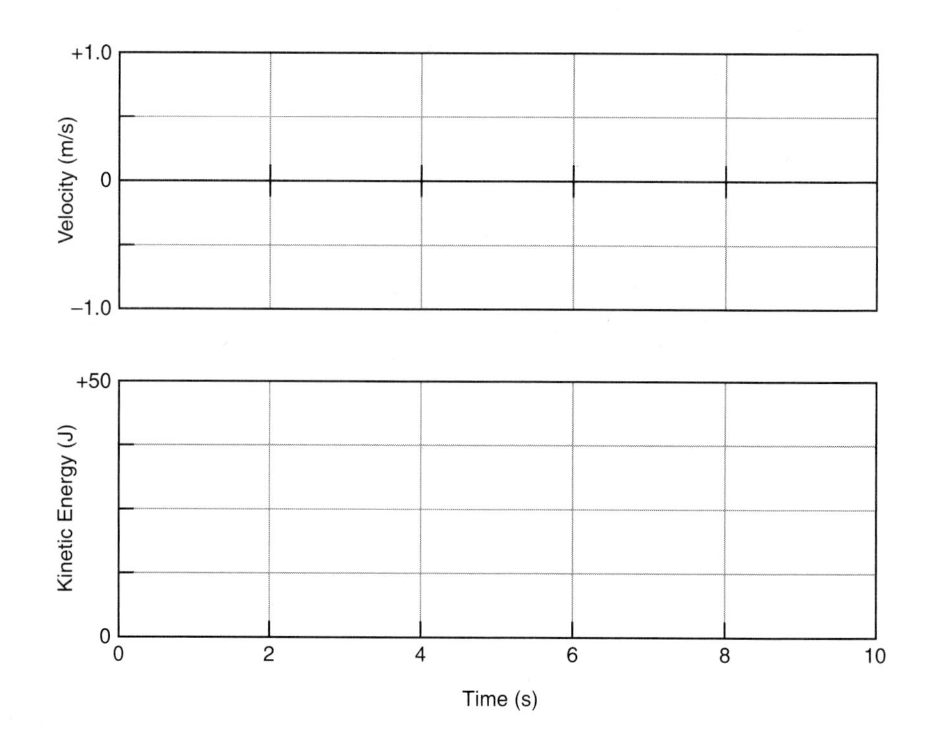

Question 3-5: In what ways does the kinetic energy graph differ from the velocity graph? Is it possible to have negative kinetic energy? Explain.

Question 3-6: Which would have a greater effect on the kinetic energy—doubling your velocity or doubling your mass? Explain.

When you apply a force to an object in the absence of friction, the object always accelerates. The force does work and the kinetic energy of the object increases. Clearly, there is some relationship between the work done on the object and the change in its kinetic energy.

Prediction 3-1: What do you think is the relationship between work done and change in kinetic energy of an object? Explain.

In the next activity, you will examine this relationship, called the *work–energy principle,* by doing work on a cart with a spring. You will need the following:

- computer-based laboratory system
- *RealTime Physics Mechanics* experiment configuration files
- force probe
- motion detector
- low-friction cart with flag
- support for motion detector with rod
- 500-g mass
- smooth ramp or other level surface 2–3 m long
- meter stick
- spring

Activity 3-3: Work–Energy Principle

1. Set up the ramp, cart, flag, motion detector, force probe, and spring as shown in the diagram that follows.

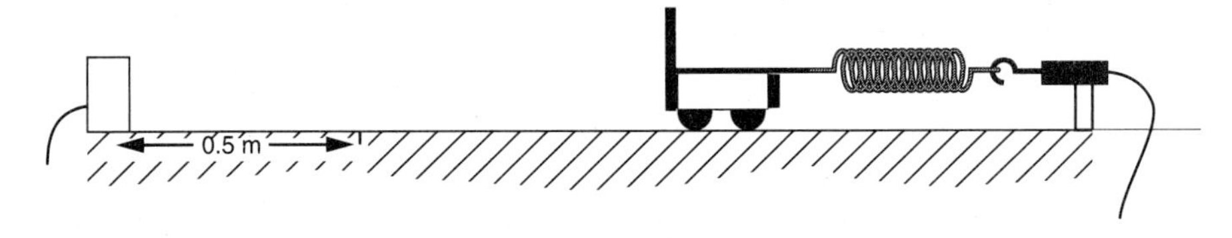

2. Open the experiment file called **Work–Energy (L11A3-3)** to display the force and kinetic energy vs. position axes that follow.

3. *Be sure that the motion detector sees the cart over the whole distance of interest—from the position where the spring is stretched about 1.0 m to the position where it is just about unstretched.*

4. Measure the mass of the cart, **and enter this value in the formula for kinetic energy.**

Mass of cart:__________kg

5. **Calibrate** the force probe using a hanging mass of 500 g (force of 4.9 N) or **load the calibration** (if it hasn't already been calibrated in Investigation 1.) (If you are using a Hall effect force probe, the spacing between the magnet and the sensor should be the same as in Investigation 1.)

6. **Zero** the force probe with the spring hanging loosely. Then pull the cart along the track so that the spring is stretched about 1.0 m from the unstretched position.

7. **Begin graphing,** and release the cart, allowing the spring to pull it back at least to the unstretched position. When you get a good set of graphs, sketch or **print** them.

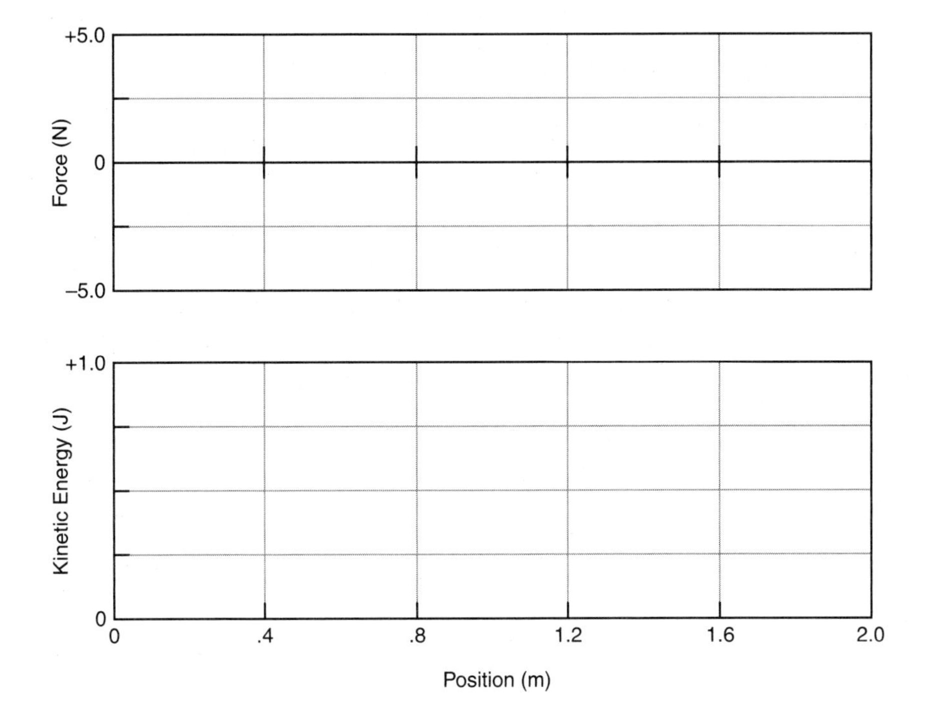

Note that the top graph displays the force applied by the spring on the cart vs. position. It is possible to find the work done by the spring force for the displacement of the cart between any two positions. This can be done by finding the area under the curve using the **integration routine** in the software, as in Activity 2-1 (and Extension 2-2).

The kinetic energy of the cart can be found directly from the bottom graph for any position of the cart.

8. Find the change in kinetic energy of the cart after it is released from the initial position (where the kinetic energy is zero) to several different final positions. Use the **analysis feature** of the software. Also find the work done by the spring up to that position.

   Record these values of work and change in kinetic energy in Table 3-3. Also determine from your graph the position of the cart where it is released and record it in the table.

Question 3-7: How does the work done on the cart by the spring compare to its change in kinetic energy? Does this agree with your prediction?

Question 3-8: State the work–energy principle that relates work to kinetic energy change in words for the cart and spring system that you have just examined.

Table 3-3

Position of cart (m)	Work done (J)	Change in kinetic energy (J)
Initial:		

Name________________________ Date____________ Partners__________________________

Homework for Lab 11: Work and Energy

1. The block below is pulled a distance of 2.50 m. How much work is done by the force? Show your work and include units.

F = 4.50 N

Answer:______

2. Now the force acts in a direction 30° above the horizontal as shown below. If the block is again moved 2.50 m, how much work is done by the force? Show your work and include units.

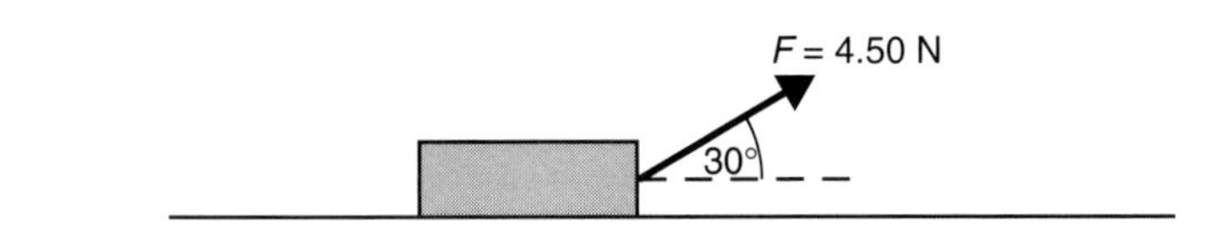

Answer:______

3. Two objects of different mass start from rest, are pulled by the same magnitude net force, and are moved through the same distance. The work done on object A is 500 J. After the force has pulled each object, object A moves twice as fast as object B. Answer the following questions and show your work.

How much work is done on object B? ______

What is the kinetic energy of object A after being pulled? ______

What is the kinetic energy of object B after being pulled? ______

What is the ratio of the mass of object A to the mass of object B? ______

4. An object of mass 0.550 kg is lifted from the floor to a height of 3.50 m at a constant speed.

   How much work is done by the lifting force (include units)?

   How much work is done by the Earth on the object?

   What is the *net* work done on the object?

   What is the change in kinetic energy of the object?

   Are your results consistent with the work-energy principle? Explain.

5. If the object in Question 4 is released from rest after it is lifted, what is its kinetic energy just before it hits the floor? What is its velocity? Show your work and include units.

   Answers: Kinetic energy:______ Velocity:______

6. A force acts on an object of mass 0.425 kg. The force varies with position as shown in the graph that follows.

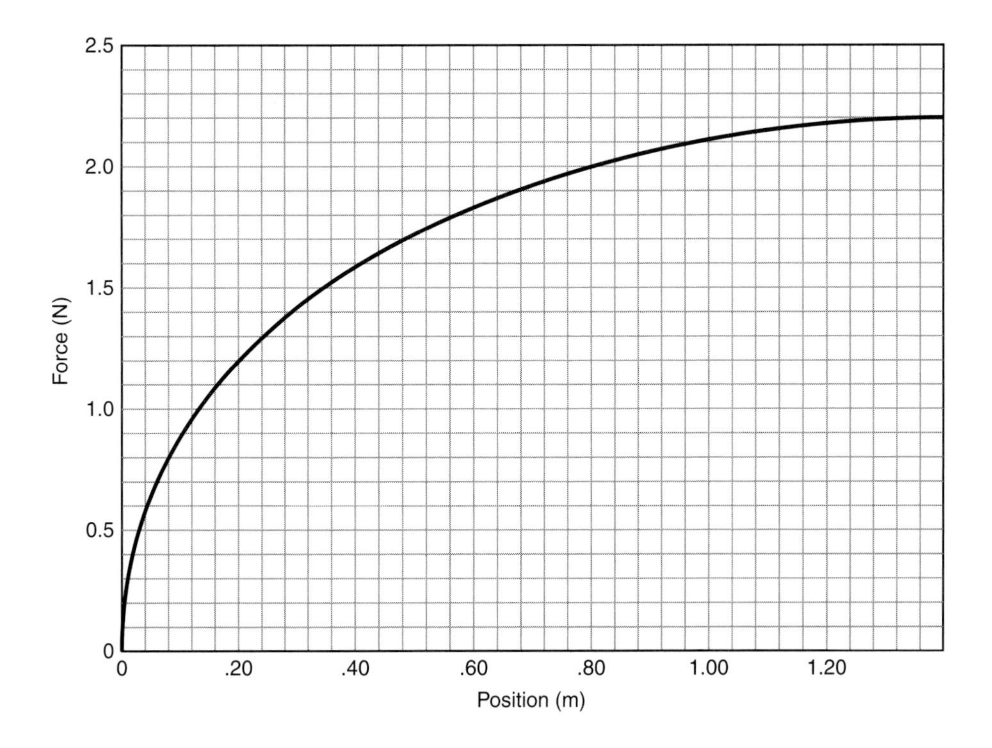

Find the work done by the force in moving the object from 0.40 m to 1.20 m. Explain your calculation and give units.

Answer:______

7. Assuming that there is no friction and that the object in Question 6 starts from rest at 0.40 m, what is the object's kinetic energy when it reaches 1.20 m? Show your calculation and give units.

Answer:______

8. What is the velocity of the object in Question 6 when it reaches 1.20 m? Show your calculation and give units.

Answer:______

Name____ Date_____________________

Pre-Lab Preparation Sheet for Lab 12: Conservation of Energy

(Due at the beginning of Lab 12)

Directions:
Read over Lab 12 and then answer the following questions about the procedures.

1. How is gravitational potential energy defined?

2. What is your Prediction 1-2? How will kinetic energy, gravitational potential energy, and mechanical energy change as the ball falls?

3. How will the height H be used in Activity 1-3?

4. How will you find the spring constant in Activity 2-1?

5. What is your Prediction 2-2? How will the kinetic energy, elastic potential energy, and mechanical energy change as the mass oscillates up and down?

Name________________________ Date____________ Partners________________________

LAB 12: CONSERVATION OF ENERGY

In order to understand the equivalence of mass and energy, we must go back to two conservation principles which . . . held a high place in pre-relativity physics. These were the principle of the conservation of energy and the principle of the conservation of mass.

—Albert Einstein

OBJECTIVES

- To understand the concept of potential energy.
- To understand the concept of mechanical energy of a system.
- To investigate situations where mechanical energy is conserved and those where it is not.

OVERVIEW

In Lab 11 on work and energy, we defined the kinetic energy associated with the motion of a rigid object:

$$K = \tfrac{1}{2}mv^2$$

where m is the mass of the object, and v is its speed. In this lab we will consider other forms of energy associated with both gravitational and spring forces.

Suppose you lift an object steadily at a slow constant velocity near the surface of the Earth so that you can ignore any change in kinetic energy. You must do work (apply a force over a distance) to lift the object because you are pulling it away from the Earth. The lifted object now has the *potential* to fall back to its original height, gaining kinetic energy as it falls. Thus, if you let the object go, it will gain kinetic energy as it falls toward the Earth.

It is very useful to define the *gravitational potential energy* of an object at height y (relative to a height $y = 0$) as *the amount of work needed to move the object away from the Earth at constant velocity through a distance y.*

If we use this definition, then the potential energy of an object is a maximum when it is at its highest point. If we let it fall, then the potential energy becomes smaller and smaller as it falls toward the Earth while the kinetic energy increases as it falls. We can now think of kinetic and potential energy to be two different forms of mechanical energy. We define the *mechanical energy* as the sum of these two energies.

Is the mechanical energy constant during the time the mass falls toward the Earth? If it is, then the amount of mechanical energy doesn't change, and we say that mechanical energy is *conserved.* If mechanical energy is conserved in other situations, we might be able to hypothesize a law of conservation of mechanical energy as follows: *In certain situations, the sum of the kinetic and potential energy, called the mechanical energy, is a constant at all times. It is conserved.*

The concept of mechanical energy conservation raises a number of questions. Does it hold quantitatively for falling masses—Is the sum of the calculated potential and kinetic energies exactly the same number as the mass falls? Can we apply a similar concept to masses experiencing other forces, such as those exerted by springs? Perhaps we can find another definition for *elastic potential energy* for a mass–spring system. In that case could we say that mechanical energy will also be conserved for an object attached to a spring? Often there are frictional forces involved with motion. Will mechanical energy be conserved for objects experiencing frictional forces, like those encountered in sliding?

In this lab you will begin by exploring the common definition of *gravitational potential energy* to see if it makes sense. You will then measure the *mechanical energy,* defined as the sum of gravitational potential energy and kinetic energy, to see if it is conserved when the gravitational force is the only force acting. Next, you will explore a system where the only net force is exerted by a spring and see the definition of *elastic potential energy.* You will measure the mechanical energy of this system and see if it is conserved. Finally, you will explore what effects sliding frictional forces or air resistance forces have on systems. You will explore whether or not mechanical energy is still conserved in such systems.

INVESTIGATION 1: GRAVITATIONAL POTENTIAL ENERGY

Suppose that an object of mass m is lifted slowly through a distance y. To cause the object to move upward at a constant velocity, you will need to apply a constant force upward just equal to the gravitational force, which is downward.

Question 1-1: Based on your knowledge from Lab 11, how much work will you do to lift the object through a distance y? Explain.

We choose to define the gravitational potential energy, U^{grav}, of an object of mass m to be equal to the work done against the gravitational force to lift it:

$$U^{grav} = mgy$$

Now you can use this equation to calculate the potential energy of a ball. You will need the following:

- computer-based laboratory system
- *RealTime Physics Mechanics* experiment configuration files
- motion detector
- large rubber ball (e.g., a basketball)
- screen to protect the motion detector

Activity 1-1: Measuring Potential Energy

1. Set up the motion detector on the floor with the screen on top of it.
2. Open the experiment file called **Measuring Grav. Pot. E. (L12A1-1).**
3. Measure the mass of the ball:______kg
4. **Configure the software with a new column** called Gravitational Potential (U^{grav}) calculated from mass times gravitational acceleration, g times the position measured by the motion detector. Then gravitational potential energy will be graphed in real time. **Display velocity on the top and kinetic energy on the bottom,** as shown on the axes that follow.
5. Hold the ball from the sides or above with your hands. Starting with the ball about 0.5 m above the motion detector (*keeping your hands and body out of the way of the motion detector*), **begin graphing** position and gravitational potential energy as you raise the ball to about 2.0 m above the motion detector. *Be sure that the ball remains directly above the motion detector.*
6. Sketch your results below or **print** your graphs and affix them over the axes.

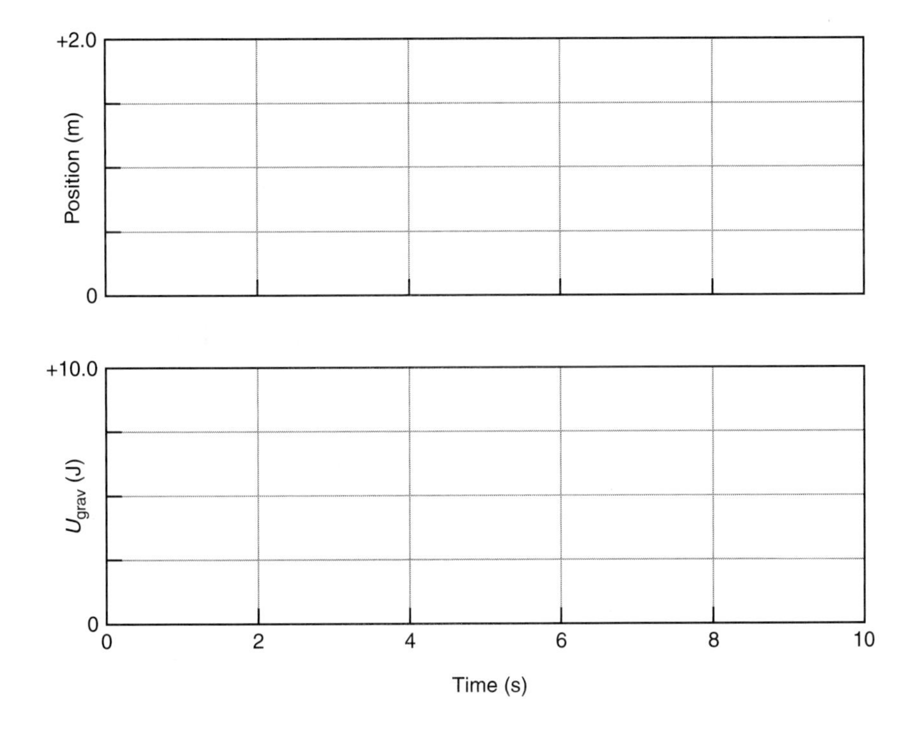

Question 1-2: Do the two graphs look similar? Does this surprise you? Explain.

Question 1-3: Gravitational potential energy is always measured with respect to a particular height where its value is defined to be zero. In this case what has been chosen as this reference level? In other words, for what location of the ball would its gravitational potential energy be zero?

Question 1-4: Suppose that the ball is dropped and you know its velocity at a certain time. What equation would you use to calculate the kinetic energy of the ball?

Prediction 1-1: Suppose that the ball is dropped from some height. What equation would you use to calculate the mechanical energy (the sum of the gravitational potential energy and the kinetic energy)?

Prediction 1-2: As the ball falls, how will the kinetic energy change? How will the gravitational potential energy change? How will the mechanical energy change?

Activity 1-2: Mechanical Energy

We can check the last predictions by measuring the two types of mechanical energy and their sum as the ball falls.

1. Open the experiment file called **Mech. Energy (L12A1-2)** to display the axes for K, U^{grav}, and mechanical energy vs. position that follow.

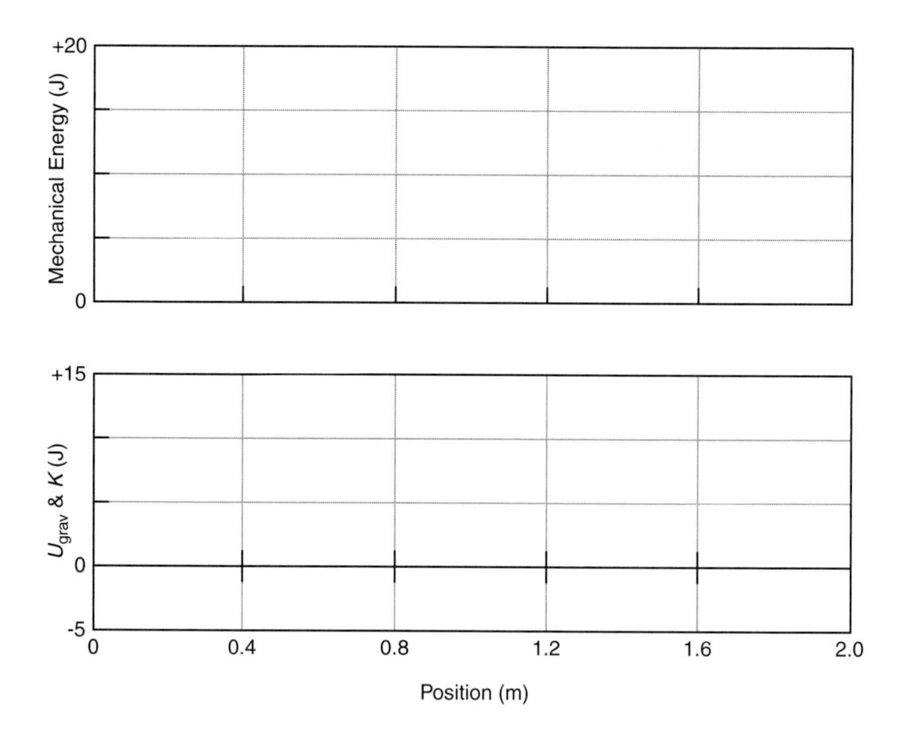

2. You will need to **enter** the mass of the ball (which you measured in the previous activity) into the formulas in the software for kinetic energy and gravitational potential energy, so that these quantities are calculated correctly.

   The mechanical energy is calculated as $U^{grav} + K$.

3. You are now ready to examine how kinetic energy, gravitational potential energy and mechanical energy vary as the ball drops. Hold the ball about 2 m

directly above the motion detector. *Be sure that it will fall on a straight path directly toward the motion detector.*

4. **Begin graphing,** and *as soon as the motion detector starts clicking,* release the ball. *Be sure that your body and hands are out of the path of the motion detector after the ball is released.*

5. Sketch your results on the previous axes or **print** your graphs and affix them over the axes.

6. Label the interval during which the ball was falling with arrows at the beginning and end.

Question 1-5: How did the variation of kinetic energy and gravitational potential energy compare to your predictions?

Question 1-6: What seems to be true about the mechanical energy defined as the sum of the kinetic energy and the gravitational potential energy? Did this agree with your prediction?

Another system where the gravitational force is essentially the only net force is a cart with very small friction moving on an inclined ramp. You can easily investigate the mechancial energy for this system as the cart rolls down the ramp. In addition to the equipment you have been using, you will need the following:

- *low-friction* cart with adjustable friction pad
- smooth ramp or other level surface 2–3 m long which can be inclined
- block to elevate one end of ramp
- protractor

Activity 1-3: Gravitational Potential, Kinetic, and Mechanical Energy of a Cart Moving on an Inclined Ramp

1. Set up the ramp and motion detector as shown below. The ramp should be inclined at an angle of about 15° above the horizontal. The friction pad on the cart should not be in contact with the ramp.

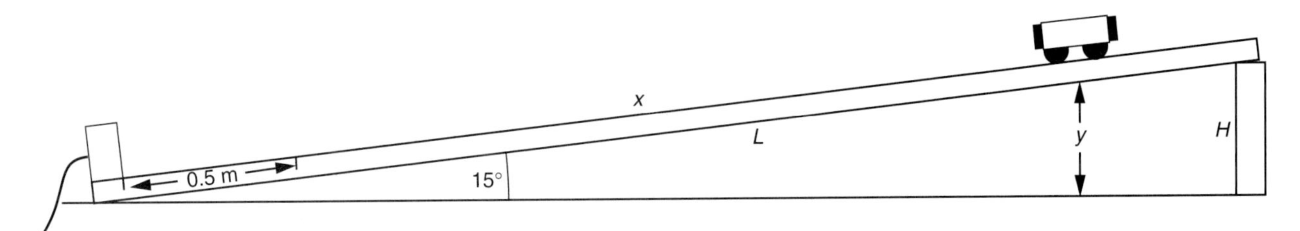

Question 1-7: The gravitational potential energy of the cart which has traveled a distance x up the ramp is mgy, where y is the height of the cart above the table top. You should be able to find an equation for U^{grav} in terms of the position x measured by the motion detector *along the ramp,* the length L of the ramp, and the elevation H of the end of the ramp. (**Hint:** $y = x \sin \theta$.)

What is the reference height for the potential energy, i.e., the height where potential energy is zero?

2. Measure the values of L and H. Also measure the mass of the cart. Write the equation for U^{grav} from Question 1-7 in the form of a single constant times x by using the values of L, H, (or for $\sin\theta$) and the gravitational acceleration, 9.80 m/s^2.

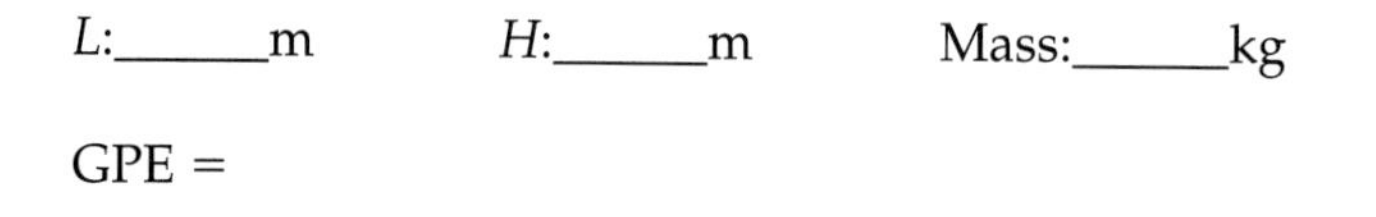

L:______m H:______m Mass:______kg

GPE =

3. Open the experiment file called **Inclined Ramp (L12A1-3)** to display the axes that follow.

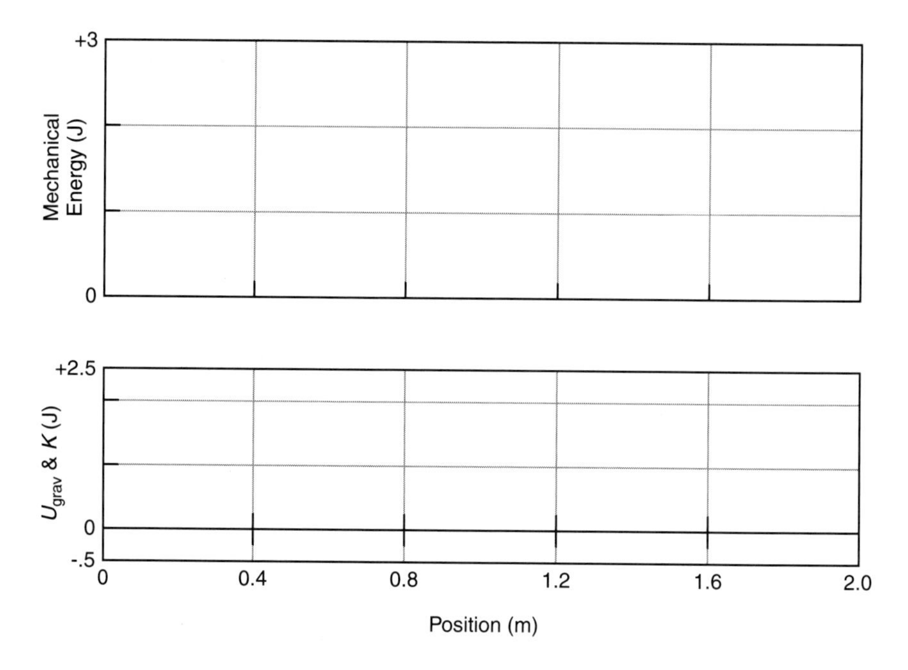

4. **Enter** the mass of the cart into the formula for kinetic energy and the constant calculated above into the formula for gravitational potential energy in the software.

   Notice that mechanical energy is calculated as $U^{grav} + K$.

Prediction 1-3: As the cart rolls down the ramp, how will the kinetic energy change? How will the gravitational potential energy change? How will the mechanical energy change?

5. *Be sure that the motion detector "sees" the cart all the way along the ramp.*
6. Hold the cart at the top of the ramp, and **begin graphing.** When you hear the clicks of the motion detector, release the cart, and stop it when it is about 0.5 m away from the motion detector.
7. Sketch your graphs on the axes or **print** your graphs and affix them over the axes.

Question 1-8: Compare your graphs to those for the falling ball in Activity 1-2. How are they similar and how are they different?

Question 1-9: What kind of variation is there in the mechanical energy as the cart rolls down the ramp? Does this agree with your prediction? Explain.

> **Comment:** The mechanical energy, the sum of the kinetic energy and gravitational potential energy, is said to be *conserved* for an object moving only under the influence of the gravitational force. That is, the mechanical energy remains constant throughout the motion of the object. This is known as the *conservation of mechanical energy.*

Prediction 1-4: Suppose that the cart is given a push up the ramp and released. It moves up, reverses direction, and comes back down again. How will the kinetic energy change? How will the gravitational potential energy change? How will the mechanical energy change? Describe in words and sketch your predictions with labeled dashed lines on the axes below.

Test your predictions.

8. Hold the cart at least 0.5 m away from the motion detector and **begin graphing.** (*Do not put your hand between the cart and the motion detector.*) When you hear the clicks of the motion detector, give the cart a push up the ramp. Stop the cart when it comes down again to about 0.5 m away from the motion detector.

9. Sketch your results on the axes that follow, or **print** your graphs and affix them over the axes.

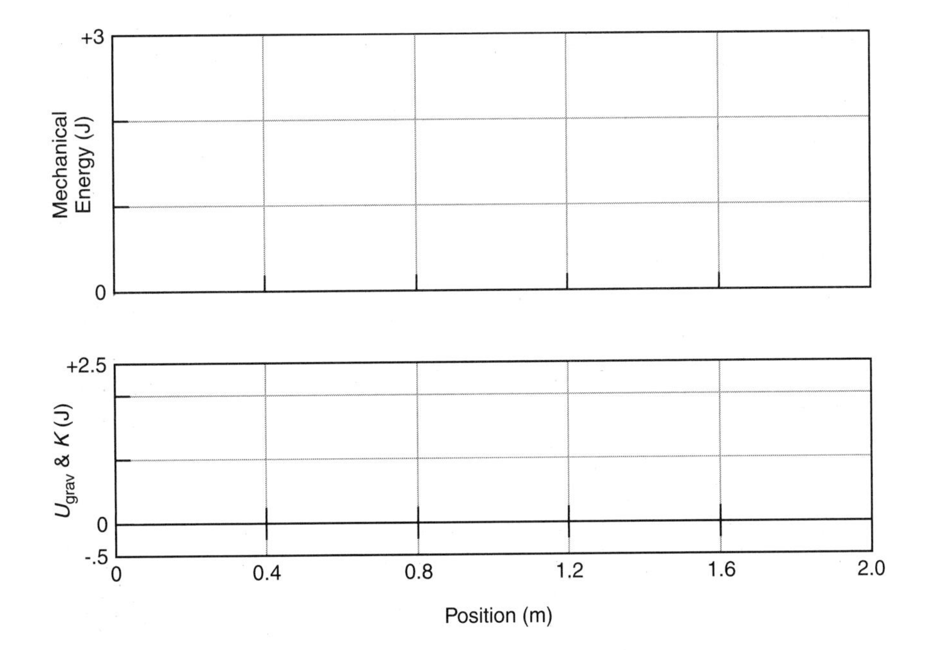

Question 1-10: How does the mechanical energy change as the cart rolls up and down the ramp? Does this agree with your prediction? Explain.

If you have additional time, do the Extension below to examine the effect of friction.

Extension 1-4: Mechanical Energy and Friction

Prediction E1-5: Suppose that there is also a frictional force acting on the cart in addition to the gravitational force. Then as the cart rolls down the ramp, how will the kinetic energy change? How will the gravitational potential energy change? How will the mechanical energy change? Compare your predictions to the case you just examined where the friction was very small.

Test your predictions. Is mechanical energy conserved when there is friction?

1. Adjust the friction pad so that there is a significant amount of friction between the pad and the ramp, but so that the cart still rolls down the ramp when released.
2. Using exactly the same setup as before, graph K, U^{grav}, and mechanical energy as the cart rolls down the ramp.
3. Sketch your graphs, or **print** and affix them over the axes that follow.

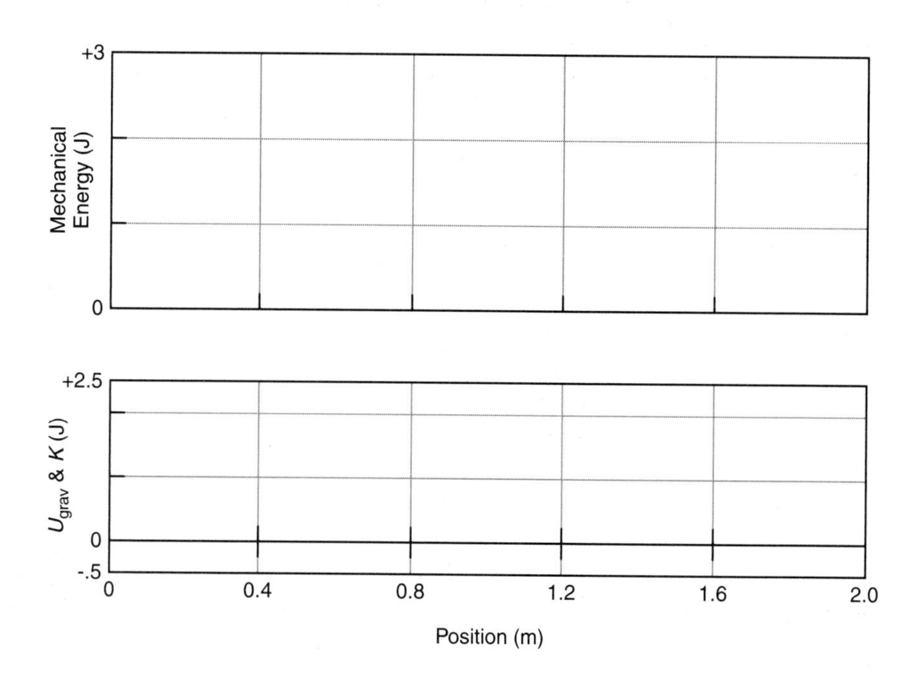

Question E1-11: Is the mechanical energy constant for the motion of the cart down the ramp *with friction*? In other words, is mechanical energy conserved?

Question E1-12: If you found that the mechanical energy was not conserved as the cart rolled down the ramp, explain what happened to the missing energy.

INVESTIGATION 2: ELASTIC POTENTIAL ENERGY

As mentioned in the Overview, it is useful to define other kinds of potential energy besides gravitational potential energy. In this investigation you will look at another common type, the *elastic potential energy,* which is associated with the elastic force exerted by a spring that obeys Hooke's law.

You have seen in Lab 11 that the magnitude of the force applied by most springs is proportional to the amount the spring is stretched from beyond its unstretched length. This is usually written $F = kx$, where k is called the spring constant.

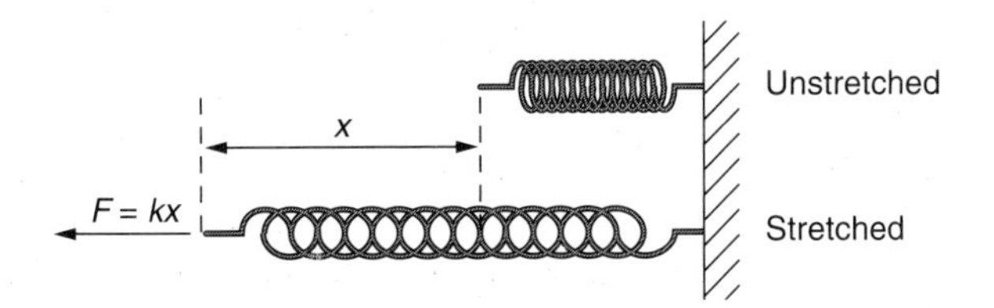

The spring constant can be measured by applying measured forces to the spring and measuring its extension.

You also saw in Lab 11 that the work done by a force can be calculated from the area under the force vs. position graph. Shown below is a force vs. position graph for a spring. Note that k is the *slope* of this graph.

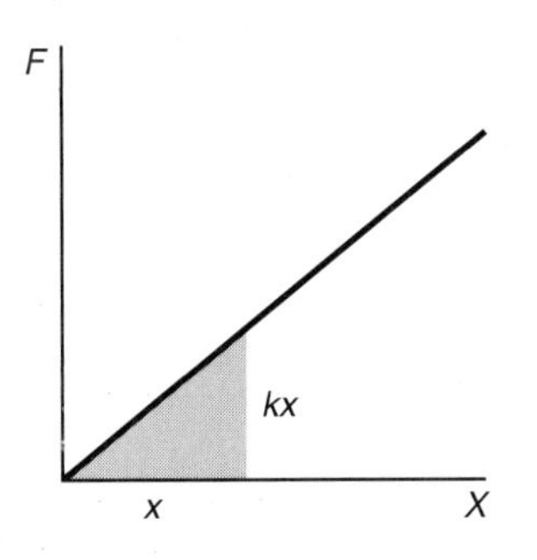

Question 2-1: How much work is done in stretching a spring of spring constant k from its unstretched length by a distance x? (**Hint:** Look at the triangle on the force vs. distance graph above and remember how you calculated the work done by a changing force in Lab 11.)

If we define the *elastic potential energy* of a spring to be the work done in stretching the spring, the definition will be analogous to the way we defined U^{grav}. In this case, the *elastic potential energy* would be

$$U^{\text{elas}} = \tfrac{1}{2}kx^2$$

In this investigation, you will measure the kinetic energy, elastic potential energy, and mechanical energy (defined as the sum of kinetic energy and elastic potential energy) of a mass hanging from a spring when air resistance is very small, and again when it is significant. You will see if the mechanical energy is conserved.

> **Note:** In the activities that follow you will explore the mechanical energy of a hanging mass oscillating on a spring. The equilibrium position depends on the gravitational force on the mass. However, it can be shown mathematically that the motion of the mass relative to the equilibrium position is only influenced by the spring force and not by the gravitational force. Therefore only U^{elas} (and not U^{grav}) needs to be included in the mechanical energy.

You will need the following:

- computer-based laboratory system
- *RealTime Physics Mechanics* experiment configuration files
- force probe
- motion detector
- mass carrier and three 50-g masses
- 200-g hanging mass
- spring
- supports to suspend the force probe and spring
- screen to protect the motion detector

Activity 2-1: Spring Constant

To calculate the elastic potential energy of a stretched spring, you need first to determine the spring constant k. Since $F = kx$, this can be found by measuring a series of forces F and the corresponding spring stretches x.

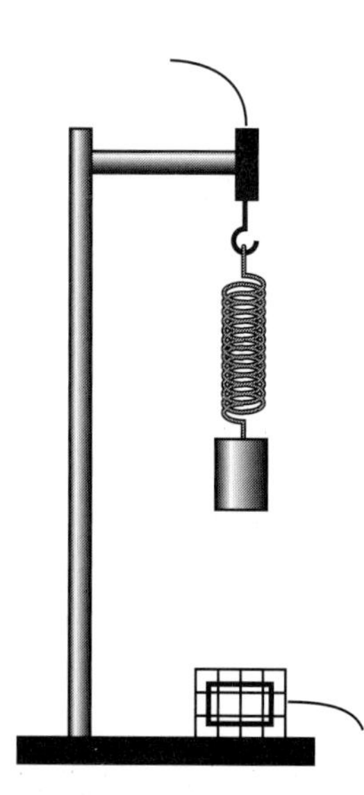

1. Set up the force probe, motion detector, and screen as shown on the left.
2. Open the experiment file called **Spring Constant (L12A2-1)** to display the axes that follow. The software is set up in **prompted event mode.** When you begin graphing, force data will be displayed continuously. After you decide to **keep** a force value, the software will allow you to enter the value of the hanging mass.
3. **Calibrate** the force probe using the 200-g mass (1.96-N weight) or **load the calibration.** (If you are using a Hall effect force probe, you may need to adjust the spacing and **check the sensitivity.**)
4. Hang the spring and the mass carrier from the force probe and then **zero** the force probe.
5. **Begin graphing. Keep** the force value, and **enter the value** of the hanging mass. Then add 50 g to the carrier, and again record this data point. Repeat until you have added 200 g to the carrier. Then **stop** graphing.

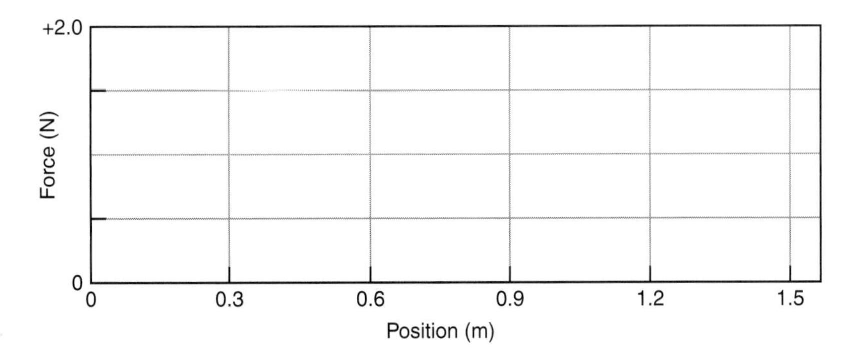

6. Use the **fit routine** in the software to find the line that fits your data, and determine the spring constant from the fit equation.

   $k =$ ______N/m

7. Sketch your graph on the axes above, or **print** your graph and affix it over the axes.

Question 2-2: Was the force exerted by the spring proportional to the displacement of the spring?

Question 2-3: What kind of a spring would have a large spring constant (large value of k)? A small spring constant?

Prediction 2-1: Hang the 200-g mass from the spring. Start it oscillating by pulling it down a small distance and letting it go. As the mass oscillates up and down, what equation would give the mechanical energy, including elastic potential energy and kinetic energy? (Note that since the mass oscillates up and down around its new equilibrium position, which is where the spring force just balances out the gravitational force [weight of the mass], you don't need to consider gravitational potential energy.)

Prediction 2-2: As the mass oscillates up and down, how will the kinetic energy change? How will the elastic potential energy change? How will the mechanical energy change?

Test your predictions.

Activity 2-2: Mechanical Energy Too

You will first need to find the new equilibrium position of the spring with the 200-g mass hanging from it.

1. Suspend the 200-g mass from the spring and be sure it is at rest.

2. Set up the axes to display Position vs. Time. **Begin graphing,** and determine the equilibrium position from the motion detector using the **analysis and statistics features** in the software.

   Equilibrium position:______m

3. Open the experiment file called **Mechanical Energy Too (L12A2-2)** to display axes like those shown below.

4. **Enter the values** of the hanging mass into the formula for kinetic energy and the spring constant and equilibrium position into the formula for elastic potential energy. Notice that mechanical energy is now calculated as $U^{\text{elas}} + K$.

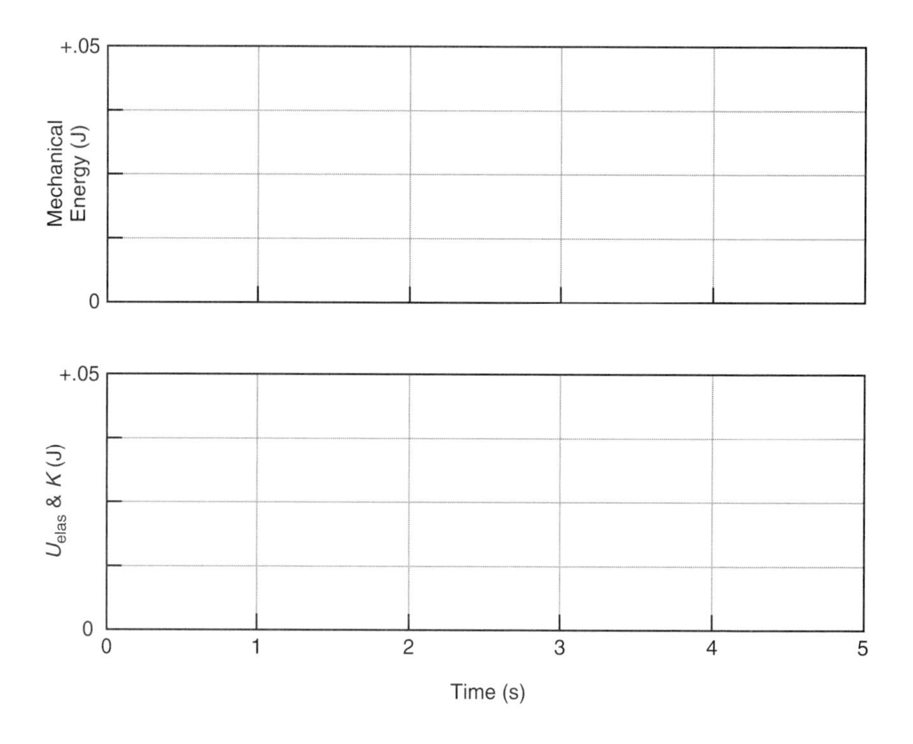

5. Start the mass oscillating with an amplitude of about 5 cm, and then **begin graphing.** Sketch your results on the axes above or **print** your graphs and affix them over the axes.

Question 2-4: How did the variations of kinetic energy and elastic potential energy compare to your predictions?

Question 2-5: What seems to be true about the mechanical energy defined as the sum of kinetic energy and the elastic potential energy? Did this agree with your prediction?

If you have additional time, do the following Extension to examine the effect of air resistance on the mechanical energy.

Extension 2-3: Mechanical Energy With Air Resistance

Prediction E2-3: Suppose that there is also significant air resistance in addition to the spring force acting on the mass as it oscillates up and down. How will the elastic potential energy change? How will the mechanical energy change? Compare your predictions to the case you just examined where the air resistance was very small.

To test your predictions you will need the following in addition to the equipment from Activity 2-2:

- large cardboard sail

1. Measure the mass of the sail. Mass of sail:______kg

   Total of hanging mass and sail:______kg

2. Attach the cardboard sail securely to the bottom of the 200-g mass, as shown in the diagram.
3. Determine the new equilibrium position.

   Equilibrium position:______m

4. **Enter** these new values into the kinetic energy and elastic potential energy formulas.
5. Start the mass and sail oscillating, and **begin graphing.** Sketch your results on the axes that follow, or **print** the results and affix the graphs over the axes.

Question E2-6: Is the mechanical energy constant for the motion of the mass *with air resistance*? Is mechanical energy conserved?

Question E2-7: If you found mechanical energy was not conserved, where did the energy go?

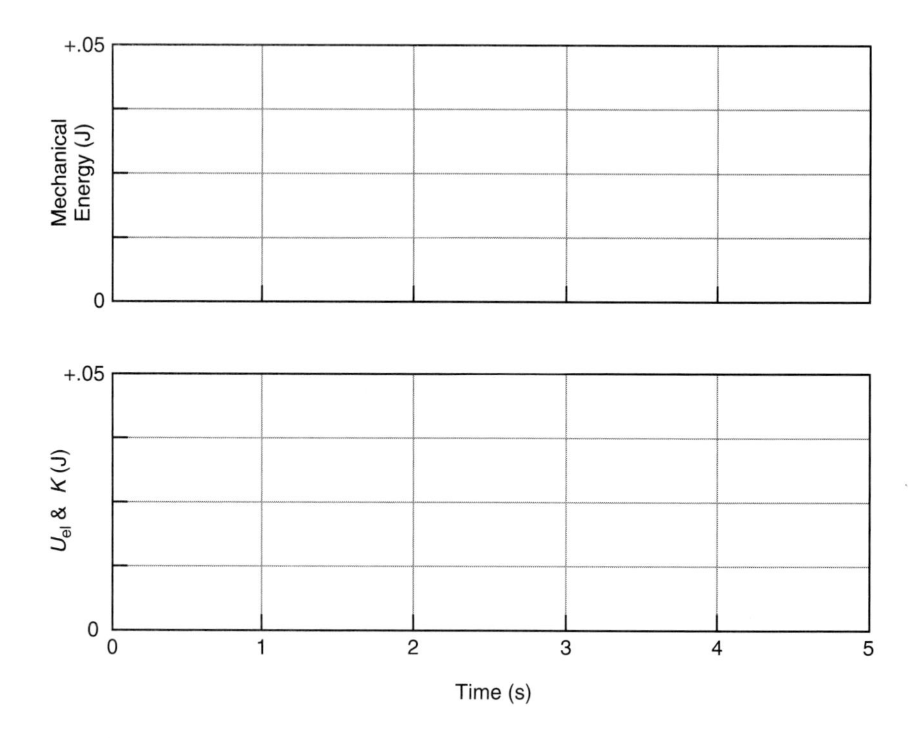

Name________________________ Date____________ Partners________________________

HOMEWORK FOR LAB 12: CONSERVATION OF ENERGY

1. A ball of mass 5.0 kg is lifted off the floor a distance of 1.7 m. What is the change in the gravitational potential energy of the ball? Show your calculation.

Answer:______ J

2. Now the ball is released from rest and falls to the floor. What is the kinetic energy of the ball just before it hits the floor? What is its velocity? Show your calculations, and explain your answers.

Answers: kinetic energy:______ J velocity:______ ms

3. Suppose that the ball is dropped from the same height as in Question 2, but is attached to a parachute. Compare the kinetic energy just before the ball hits the floor to your answer in Question 2. Is it the same, larger, or smaller? Is mechanical energy conserved in this case? Explain your answer.

4. A ball is tossed in the air and released. It moves up, reverses direction, falls back down again, and is caught at the same height it was released.

a. Considering the time interval after the ball is released and before it is caught, when does the gravitational potential energy of the ball have its maximum value? Minimum value? Explain.

b. When does the kinetic energy of the ball have its maximum value? Minimum value? Explain.

c. What about the mechanical energy of the ball? What can you say about its value at the locations described in your answers to (a) and (b)?

5. A car of mass 1000 kg is at the top of a 10° hill as shown.

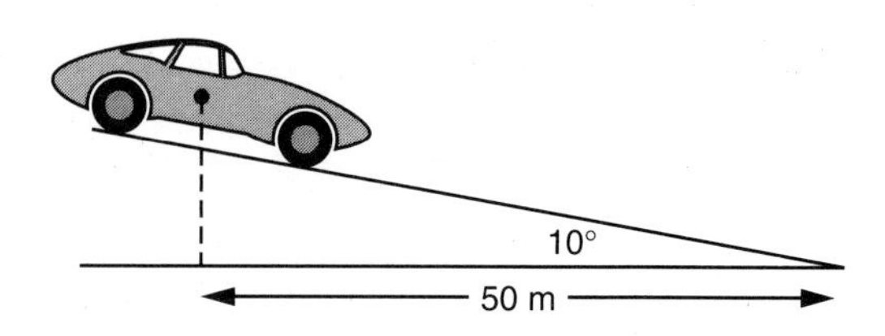

   a. What is its gravitational potential energy relative to the bottom of the hill?

   b. If the car rolls down the hill (with the engine off) with negligible friction and air resistance, what will its kinetic energy be when it reaches the bottom?

   c. Suppose instead that the amount of work done on the car by the frictional and air resistance forces as the car rolls down the hill is 50,000 J. What then is the kinetic energy of the car when it reaches the bottom of the hill?

6. The graph on the right is of the force exerted by a spring as a function of the distance it is stretched from its unstretched length. Show calculations for each part below.

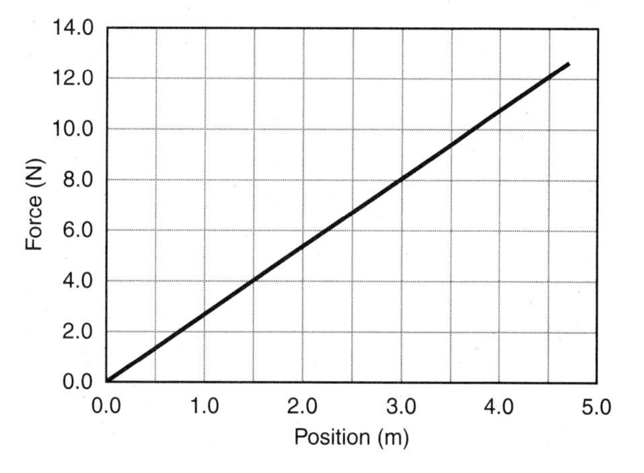

   a. What is the spring constant of this spring?

   b. What is the elastic potential energy if the spring is stretched 1.5 m from its equilibrium length?

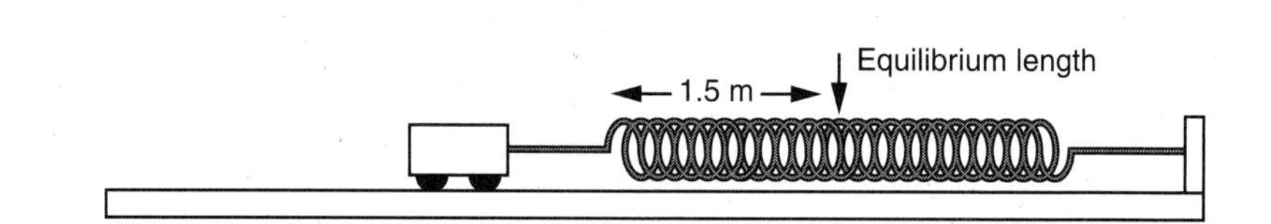

   c. A low-friction cart of mass 5.0 kg is attached to the spring, the spring is stretched 1.5 m from its equilibrium length, and then the cart is released. What is the kinetic energy of the cart just as the spring passes through its equilibrium length? What is the velocity of the cart at this position?

Appendix A: RealTime Physics Mechanics Experiment Configuration Files

Listed below are the settings in the *Experiment Configuration Files* used in these labs. These files are available from Vernier Software and Technology for the compatible software which they sell (*Logger Pro*) and from PASCO Scientific for *Data Studio*. They are listed here so that the user can set up files for any compatible hardware and software package.

Experiment File	Description	Data Collection	Data Handling	Analysis	Display
Distance (LO1A1-1a)	Graphs distance vs. time	Motion sensor 20 points/s	NA	NA	One set of graph axes with line Distance: 0–4 m Time 0–10 s
Away and Back (LO1A1-1)	Graphs position vs. time	Motion sensor 20 points/s	NA	NA	One set of graph axes with line Position: 0–2 m Time 0–15 s
Position Match (LO1A1-2)	Graphs position vs. time with preset persistent position-time graph to match	Motion sensor 20 points/s	Position Match graph is persistently displayed on the screen for comparison with collected data	NA	One set of graph axes with line Position: 0–4 m Time 0–20 s Persistent graph as in Lab 1 text
Velocity Graphs (LO1A2-1)	Graphs velocity vs. time	Motion sensor 20 points/s	NA	NA	One set of graph axes with line Velocity: −1 to +1 m/s Time 0–5 s
Velocity Match (LO1A2-2)	Graphs velocity vs. time with preset persistent velocity-time graph to match	Motion sensor 20 points/s	Velocity Match graph is persistently displayed on the screen for comparison with collected data	NA	One set of graph axes with line Velocity: −1 to +1 m/s Time 0–20 s Persistent graph as in Lab 1 text
Velocity from Position (LO1A3-1)	Graphs position and velocity vs. time	Motion sensor 20 points/s	NA	Use analysis feature, statistics feature and graphical fit routine to find average velocity from position-time and velocity-time graphs	Two sets of graph axes with lines Position 0–4 m Velocity: −2 to +2 m/s Time 0–5 s

Experiment File	Description	Data Collection	Data Handling	Analysis	Display
Constant Velocity (L01A4-1)	Graphs position and velocity vs. time	20 points/s Motion sensor	NA	NA	Two sets of graph axes with lines Position: 0–2 m Velocity: −1 to +1 m/s Time 0–5 s
Speeding Up (L02A1-1)	Graphs position, velocity and acceleration vs. time	20 points/s Motion sensor	Use to display data from first part persistently on the screen for comparison with data from second part	Use analysis feature, statistics feature, and graphical fit routine to find average acceleration from velocity-time and acceleration-time graphs	Three sets of graph axes with lines Position: 0–2 m Velocity: −1 to +1 m/s Accel.: −2 to +2 m/s^2 Time 0–3 s
Slowing Down (L02A3-1)	Graphs velocity and acceleration vs. time	20 points/s Motion sensor	Use to display data from first part persistently on the screen for comparison with data from second part	NA	Two sets of graph axes with lines Velocity: −1 to +1 m/s Accel.: −2 to +2 m/s^2 Time 0–5 s
Measuring Force (L03A1-2a)	Graphs force vs. time	20 points/s Force sensor	NA	Use analysis feature to read data from graph	One set of graph axes with line Force −10 to +10 N Time 0–10 s
Rubber Bands (L03A1-1b)	Graphs force vs. number of rubber bands pulled. Force data are continuously measured and can be kept as desired	Force sensor Event mode—keep force values and manually enter the number of rubber bands	NA	Use analysis feature and fit routine to find relationship between force and number of rubber bands pulled	One set of graph axes with line Force −10 to +10 N # Rubber bands: 0–5
Calibrating Force (L03A1-4)	Graphs force vs. time	20 points/s Force sensor	NA	Use analysis feature to read data from graph	One set of axes with line Force: −10 to +10 N Time 0–10 s

Experiment File	Description	Data Collection	Data Handling	Analysis	Display
Motion and Force (LO3A2-1)	Graphs velocity, force and acceleration vs. time	20 points/s Motion sensor and force sensor	NA	NA	Three sets of graph axes with lines Velocity: −1 to +1 m/s Force: −2 to +2 N Accel.: −2 to +2 m/s^2 Time 0–5 s
Speeding Up Again (LO3A2-2)	Graphs velocity, acceleration and force vs. time	20 points/s Motion sensor and force sensor	Use to display data from first part persistently on the screen for comparison with data from second part	Use analysis feature and statistics feature to find average force and average acceleration from graph	Three sets of graph axes with lines Velocity: −2 to +2 m/s Accel.: −2 to +2 m/s^2 Force: −2 to +2 N Time 0–3 s
Acceleration vs. Force (LO3A2-5)	Graphs avg. acceleration vs. avg. force. Data are entered manually	NA	NA	Use graphical fit routine to find relationship between average force and average acceleration	Table for entering data manually. One set of graph axes with line. Acceleration and force ranges as appropriate
Slowing Down Again (LO4A1-1)	Graphs velocity, acceleration and force vs. time	20 points/s Motion sensor and force sensor Push is positive	Use to display data from first part persistently on the screen for comparison with data from second part	NA	Three sets of graph axes with lines Velocity: −2 to +2 m/s Accel.: −2 to +2 m/s^2 Force: −2 to +2 N Time 0–3 s
Cooperating Fan Units (LO4A2-2a)	Graphs force vs. time	20 points/s Force sensor	NA	Use analysis feature and statistics feature to find average force from graph	One set of graph axes with line Force: −2 to +2 N Time 0–10 s
Velocity and Acceleration (LO4A2-2b)	Graphs velocity and acceleration vs. time	20 points/s Motion sensor	Use to display data from first part persistently on the screen for comparison with data from second part	Use analysis feature and statistics feature to find average accelerations from graph	Two sets of graph axes with lines Velocity: −2 to +2 m/s Accel.: −2 to +2 m/s^2 Time 0–5 s

Experiment File	Description	Data Collection	Data Handling	Analysis	Display
Dueling Fan Units (L04A3-1)	Graphs velocity and acceleration vs. time	20 points/s Motion sensor	Use to display data from first part persistently on the screen for comparison with data from second part	NA	Two sets of graph axes with lines Velocity: -2 to $+2$ m/s Accel.: -2 to $+2$ m/s^2 Time 0–5 s
Once a Pull (L04E3-2)	Graphs velocity, acceleration and force vs. time	20 points/s Motion sensor and force sensor	NA	NA	Three sets of graph axes with lines Velocity: -2 to $+2$ m/s Accel.: -2 to $+2$ m/s^2 Force: -2 to $+2$ N Time 0–5 s
Acceleration and Mass (L05A1-1)	Graphs velocity, acceleration and force vs. time	20 points/s Motion sensor and force sensor	Use to display data from first part persistently on the screen for comparison with data from second part	Use analysis feature and statistics feature to find average forces and average accelerations from graphs	Three sets of graph axes with lines Velocity: -2 to $+2$ m/s Accel.: -2 to $+2$ m/s^2 Force: -2 to $+2$ N Time 0–3 s
Avg. Acceleration vs. Mass (L05A1-4)	Graphs avg. acceleration vs. mass. Data are entered manually	NA	NA	Use graphical fit routine to find relationship between average acceleration and mass	Table for entering data manually. One set of graph axes with line. Acceleration and mass ranges as appropriate.
Acceleration with 1.0 N (L05A2-2)	Graphs velocity, acceleration and force vs. time	20 points/s Motion sensor and force sensor	NA	Use analysis feature and statistics feature to find average forces and average accelerations from graphs	Three sets of graph axes with lines Velocity: -2 to $+2$ m/s Accel.: -2 to $+2$ m/s^2 Force: -2 to $+2$ N Time 0–3 s
Falling Ball (L06A1-1)	Graphs velocity and acceleration vs. time	30 points/s Motion sensor Distance away is negative	Use to display data from first part persistently on the screen for comparison with data from second part	Use analysis and statistics features to find average acceleration from graph	Two sets of graph axes with lines Velocity: -5 to $+5$ m/s Accel.: -15 to $+15$ m/s^2 Time 0–2 s

Experiment File	Description	Data Collection	Data Handling	Analysis	Display
Inclined Ramp (LO6A1-5)	Graphs velocity and acceleration vs. time	20 points/s Motion sensor Distance away is negative	NA	Use analysis and statistics features to find average acceleration from graph	Two sets of graph axes with lines Velocity: −2 to +2 m/s Accel.: −4 to +4 m/s^2 Time 0–5 s
Action of Friction (LO7A1-1)	Graphs velocity, acceleration and force vs. time	20 points/s Motion sensor and force sensor	Use to display data from first part persistently on the screen for comparison with data from second part	Use analysis feature and statistics feature to find average forces and average accelerations from graphs	Three sets of graph axes with lines Velocity: −2 to +2 m/s Accel.: −2 to +2 m/s^2 Force: −2 to +2 N Time 0–5 s
Static and Kinetic Friction (LO7E1-2)	Graphs force and velocity vs. time	20 points/s Motion sensor and force sensor	Use to display data from first part persistently on the screen for comparison with data from second part	Use analysis and statistics features to find average forces from graph	Two sets of graph axes with lines Force: −2 to +2 m/s Velocity: −0.5 to +0.5 m/s Time 0–5 s
Tug-of-War (LO7A2-1)	Graphs force sensors 1 and 2 vs. time	20 points/s Two force sensors Sign of force sensor 2 is reversed	NA	NA	Two sets of graph axes with lines Forces: −10 to +10 N Time: 0–10 s
Tension Forces (LO7A3-1)	Graphs force sensors 1 and 2 vs. time	20 points/s Two force sensors Sign of force sensor 2 is reversed	NA	NA	Two sets of graph axes with lines Forces: −10 to +10 N Time: 0–10 s
Tension and N2 (LO7E3-3)	Graphs force and acceleration vs. time	20 points/s Motion sensor and force sensor	NA	Use analysis feature and statistics feature to find average forces and average accelerations from graphs	Two sets of graph axes with lines Force: −5 to +5 N Accel.: −4 to +4 m/s^2 Time 0–3 s
Clay vs. Superball (LO8A1-2)	Graphs force vs. time	5000 points/s Force sensor only Push is positive Triggered mode	Use to display data from first part persistently on the screen for comparison with data from second part	Use analysis feature to find maximum forces from graphs	One set of graph axes with line Force: can be uncalibrated Time 0–0.2 s

Experiment File	Description	Data Collection	Data Handling	Analysis	Display
Impulse and Momentum (L08A2-2)	Graphs velocity and force vs. time	50 points/s Motion sensor and force sensor Velocity toward motion detector positive Push is positive	NA	Use analysis and statistics features to find the average velocities from graphs Use integration routine to find area under force vs. time graph	Two sets of graph axes with lines Velocity −2 to +2 m/s Force: −10 to +10 N Time 0–2 s
Collisions (L09A1-1)	Graphs force sensors 1 and 2 vs. time	5000 points/s Two force sensors Sign of one sensor reversed Triggered mode	NA	Use integration routine to find area under force vs. time graphs	Two sets of graph axes with lines Forces: −10 to +10 N Time 0–0.2 s
Other Interactions (L09A1-2)	Graphs force sensors 1 and 2 vs. time	20 points/s Two force sensors Sign of one sensor reversed	NA	NA	Two sets of graph axes with lines Forces: −10 to +10 N Time 0–5 s
Inelastic Collisions (L09A2-1)	Graphs velocity vs. time	20 points/s Motion sensor	NA	Use analysis and statistics features to find the velocities from graphs	One set of axes with line Velocity −2 to +2 m/s Time 0–5 s
Position vs. Time (L10A1-1)	Graphs position vs. time. Data are entered manually	NA	NA	NA	One set of graph axes with line. Position and time ranges as appropriate
Velocity vs. Time (L10A1-2)	Graphs velocity vs. time. Data are entered manually	NA	NA	Use the fit routine to find relationship between velocity and time	One set of graph axes with line. Velocity and time ranges as appropriate
2-D Position vs. Time (L10A2-1a)	Graphs x-position and y-position vs. time. Data are entered manually	NA	NA	NA	One set of graph axes with line. Position and time ranges as appropriate
2-D Velocity vs. Time (L10A-2b)	Graphs x-velocity and y-velocity vs. time. Data are entered manually	NA	NA	Use the fit routine to find relationships between the velocities and time, and to find the x and y accelerations	One set of graph axes with line. Velocity and time ranges as appropriate

Experiment File	Description	Data Collection	Data Handling	Analysis	Display
Force and Work (L11A1-2)	Graphs force vs. time	20 points/s Force sensor only	Use to display data from first part persistently on the screen for comparison with data from second part	Use analysis and statistics features to find the average forces from the graphs	One set of axes with line Force −5 to +5 N Time 0–10 s
Work in Lifting (L11A2-1)	Graphs velocity and force vs. time. Then switch to force vs. position	20 points/s Motion sensor and force sensor	NA	Use analysis and statistics features to find the average force from the graph. Use the integration routine to find the area under the force vs. position graph	Two sets of axes with line Velocity −1 to +1 m/s Force −5 to +5 N Time 0–10 s
Stretching Spring (L11E2-2)	Graphs force vs. position	20 points/s Motion sensor and force sensor Distance away from motion sensor is negative	NA	Use the integration routine to find the area under the force vs. position graph	One set of axes with line Force −5 to +5 N Position 0–2 m
Kinetic Energy (L11A3-2)	Graphs velocity and kinetic energy vs. time	20 points/s Motion sensor	Calculated column for kinetic energy with mass entered manually	NA	Two sets of axes with lines Velocity −1 to +1 m/s Kinetic energy 0 to 50 J Time 0–10 s
Work-Energy (L11A3-3)	Graphs force and kinetic energy vs. position	20 points/s Motion sensor and force probe	Calculated column for kinetic energy with mass entered manually	Use the integration routine to find the area under the force vs. position graph. Use the analysis feature to find the kinetic energy from the graph	Two sets of axes with lines Force −5 to +5 N Kinetic energy 0 to 1 J Time 0–2 s
Measuring Grav. Pot. E. (L12A1-1)	Graphs position and gravitational potential energy vs. time	20 points/s Motion sensor	Calculated column for gravitational potential energy with mass entered manually	NA	Two sets of axes with lines Position 0–2 m Gravitational potential energy 0 to 10 J Time 0–10 s

Experiment File	Description	Data Collection	Data Handling	Analysis	Display
Mechanical Energy (L12A1-2)	Graphs mechanical energy, kinetic energy and gravitational potential energy vs. position	20 points/s Motion sensor	Calculated columns for mechanical energy, kinetic energy and gravitational potential energy with mass entered manually	NA	Two sets of axes with lines Mechanical energy 0–20 J Kinetic energy and gravitational potential energy on same axes −5 to 15 J Position 0–2 m
Inclined Ramp (L12A1-3)	Graphs mechanical energy, kinetic energy and gravitational potential energy vs. position	20 points/s Motion sensor	Calculated columns for mechanical energy, kinetic energy and gravitational potential energy with mass entered manually	NA	Two sets of axes with lines Mechanical energy 0–3 J Kinetic energy and gravitational potential energy on same axes −0.5 to +2.5 J Position 0–2 m
Spring Constant (L12A2-1)	Graphs force vs. position	20 points/s Motion sensor and force sensor. Event mode—keep force and position values only when desired	New prompted manual column to enter mass	Use graphical fit routine to find relationship between force and position	One set of axes with line Force 0–2 N Position 0–1.5 m
Mechanical Energy Too (L12A2-2)	Graphs mechanical energy, kinetic energy and elastic potential energy vs. position	20 points/s Motion sensor Velocity and acceleration data averaging reduced to 3 points	Calculated columns for mechanical energy, kinetic energy and elastic potential energy with mass, spring constant and equilibrium position entered manually	Use the analysis and statistics features to find the equilibrium position from graph	Two sets of axes with lines Total mechanical energy 0–0.05 J Kinetic and elastic potential energy on same axes 0–0.05 J Time 0–5 s